SIMPLICIUS

On Aristotle's

"On the Heavens 2.1-9"

SIMPLICIUS

On Aristotle's "On the Heavens 2.1-9"

Translated by
Ian Mueller

Cornell University Press

Ithaca, New York

First published 2004 by Cornell University Press.

ISBN 0-8014-4102-1

Acknowledgments

The present translations have been made possible by generous and imaginative funding from the following sources: the National Endowment for the Humanities, Division of Research Programs, an independent federal agency of the USA; the Leverhulme Trust; the British Academy; the Jowett Copyright Trustees; the Royal Society (UK); Centro Internazionale A. Beltrame di Storia dello Spazio e del Tempo (Padua); Mario Mignucci; Liverpool University; the Leventis Foundation; the Arts and Humanities Research Board of the British Academy; the Esmée Fairbairn Charitable Trust; the Henry Brown Trust; Mr and Mrs N. Egon; the Netherlands Organisation for Scientific Research (NWO/GW). The editor wishes to thank D. Furley, C. Wildberg, B. Morison, and D. Russell for their comments, and John Sellars for preparing the volume for press.

Printed and bound in Great Britain

Librarians: Library of Congress Cataloging-in-Publication Data are available.

Contents

Preface

The first draft of this translation of Simplicius' commentary on Book 2 of Aristotle's *De Caelo* was finished in 1998-99 when I was an exchange fellow at the Centre d'Histoire des Sciences et des Philosophies Arabes et Médiévales (CHSPAM) at the Centre National de la Recherche Scientifique (CNRS) in Paris.

I would like to thank the personnel, staff, and associates of CHSPAM for their assistance, intellectual and personal, in making my year in Paris profitable and pleasurable, with particular thanks to the then director, Roshdi Rashed and to Tony Lévy, Régis Morelon, Pierre Pellegrin, Muriel Rouabah, and Bernard Vitrac for their interest and kindness. I also thank the University of Chicago and the CNRS for choosing me as an exchange fellow for that year.

Completion of the translation was set back by two years because of administrative duties at the University of Chicago. I am grateful to the readers of the translation for their comments and suggestions, which I have followed more often than not. The names of most of them are not known to me, but I am in a position to mention and especially thank Allen Bowen and David Furley. John Sellars has done admirable editorial work, making many improvements, including the elimination of all too many mistakes. For the all too many that remain I am, of course, fully responsible.

My deepest debt is, as always, to my wife Janel Mueller, who, while pursuing an academic and administrative career of her own, provided me with more support and encouragement than anyone is entitled to.

Details concerning the text translated and some of my translational practices can be found in the Introduction.

Chicago — Ian Mueller

Abbreviations

CAG = *Commentaria in Aristotelem Graeca*, Berlin: G. Reimer, 1882-1909.

DK = Diels, Hermann, and Kranz, Walther (ed. and trans.), *Die Fragmente der Vorsokratiker*, 6th edn, Berlin: Weidmann, 1954.

DSB = Gillispie, Charles Coulston (ed.), *The Dictionary of Scientific Biography*, 16 vols, New York: Charles Scribner's Sons, 1970-80.

El. = *Euclid's Elements*, vols 1-4 in Heiberg, J.L. and Menge, Hermann (ed. and trans.), *Euclidis Opera Omnia*, Leipzig: Teubner, 1883-1916.

Guthrie = Guthrie, W.K.C. (ed. and trans.), *Aristotle, On the Heavens*, Cambridge, MA: Harvard University Press, and London: William Heinemann, 1939.

Heiberg = *CAG*, vol. 7.

Karsten = Karsten, Simon (ed.), *Simplicii Commentarius in IV Libros Aristotelis De Caelo*, Utrecht: Kemink and Son, 1865.

Leggatt = Leggatt, Stuart (ed. and trans.), *Aristotle, On the Heavens I & II*, Warminster: Aris and Phillips, 1995.

Moraux = Moraux, Paul (ed. and trans.), *Aristote, du Ciel*, texte établi et traduit par Paul Moraux, Paris: Belles Lettres, 1965.

OED = *The Oxford English Dictionary*, 2nd edn, Oxford: Clarendon Press, 1989.

Oeuv. Comp. = Platon, *Oeuvres Complètes*, 27 vols, Paris: Belles Lettres, 1920-1964.

PW = *Paulys Realencyclopaedie der Classischen Altertumswissenschaft*, 51 vols, Stuttgart: J.B. Metzler, 1893-1997.

Introduction

1. The text

This translation of Simplicius' commentary on Book 2 of Aristotle's *De Caelo* (*On the Heavens*) is based on Heiberg's edition of the text printed as volume 7 of *CAG*. For the text of *De Caelo* itself I have relied on Moraux. Since it seems reasonably clear that Heiberg's edition should not be regarded as definitive[1] and the present textual situation affects my translational practices, I wish here to say a few words about Heiberg's edition. My remarks are based on Heiberg's preface to his edition (cited here by Roman numeral page) and his earlier, more detailed but slightly discrepant, report to the Berlin Academy (Heiberg (1892)). I confine my remarks to Book 2, the situation for Book 1 being significantly different.

For Heiberg the most important manuscript is:

A Mutinensis III E 8, thirteenth-fourteenth century, in the Este Library in Modena (Wartelle (1963) no. 1052)

Heiberg (1892) 71, singles out A for its correctness and purity. But he admits that it is badly deficient and hastily written, with frequent incorrect divisions of words, misunderstandings of abbreviations, arbitrary use of accents and breathing marks, extremely many omissions, and frequent insertions in a wrong place of words occurring in the vicinity. A glance at the apparatus on almost any page of the edition of Book 2 makes clear how often Heiberg feels forced to depart from A. On the whole these departures seem justified, but, as I shall discuss further below, there are many cases where he follows A and produces a text which seems to me impossible or at least very difficult.

Heiberg thought that A and another text, which he calls B, derived independently from a lost archetype. B stops in Book 1, the remaining pages being torn out. However, there are two other manuscripts which Heiberg took to be copied from B:

J Taurinensis C.I.3, sixteenth century, in the National Library of Turin (Wartelle (1963) no. 2086)

Perusinus cod. 51, fourteenth century, in the Municipal Library of Perugia (Wartelle (1963) no. 1658)

Heiberg makes no use of either of these manuscripts in his edition of Book 2, but it looks as though attention ought to be paid to them.[2]

Four manuscripts which Heiberg does cite are:

- C Coislinianus 169, fifteenth century, in the National Library in Paris (Wartelle (1963) no. 1560)
- D Coislinianus 166, fourteenth century, in the National Library of Paris (Wartelle (1963) no. 1558)
- E Marcianus 491, thirteenth century, in the library of San Marco, Venice (Mioni (1985) 299-300; not in Wartelle (1963))
- F Marcianus 228, fifteenth century, in the library of San Marco, Venice (Wartelle (1963) no. 2129)

Heiberg took D and E to be significantly different from A and B, and C to be intermediate between D and E, on the one hand, and A and B, on the other. C and D are, in fact, texts of *De Caelo* with extensive marginalia from Simplicius' commentary (not necessarily word-for-word quotations) rather than complete texts of Simplicius. According to Heiberg, E, which is a complete (although lacunose) text, and D were copied from the same prototype, E being copied by an uneducated scribe. E was corrected by Bessarion (E^2), using the Latin translation of William Moerbeke, a work to which I shall return shortly. Heiberg sometimes adopts readings of D or E and, less frequently, of C. Heiberg's treatment of F, which contains only Books 2 to 4, causes the greatest difficulty. Heiberg decided on quite inadequate grounds that F is a descendant of A. He cites it only where it seems useful, so that, as he says, nothing can be concluded about its contents in places where it is not mentioned in the apparatus.

Heiberg also cites three printed versions of the commentary in his apparatus:

- (a) The *editio princeps* of the Greek text: *Simplicii Commentarii in Quatuor Libros de Coelo, cum Textu Ejusdem*, Venice: Aldus Romanus and Andrea Asulani, 1526
- (b) The *editio princeps* of the Latin translation of William Moerbeke: *Simplicii philosophi acutissimi, Commentaria in Quatuor Libros De coelo Aristotelis*, Venice: Hieronymus Scotus, 1540
- (c) Karsten (1865); referred to here as Karsten

Citations of (a) are rare because Heiberg ((1892), p. 75) realised that it was a translation back into Greek of Moerbeke's Latin translation.[3] However, he did not realise that (b) was 'corrected' in the light of (a). Since Moerbeke used an older manuscript of the commentary than any now extant, it is clear that one condition of a satisfactory edition of the Greek text is a new edition of Moerbeke's translation. The situation is made even more complicated by the discovery of a translation by Robert Grosseteste of parts of Simplicius' commentary, including the whole of Book 2.[4] Clearly an edition of this text is needed for the reconstruction of Simplicius' Greek. In my reports on what is in Heiberg's apparatus criticus I usually omit what he says about (b).

Karsten's edition was published one year after his death. It includes no critical apparatus, and has no preface by Karsten. Throughout it is based on single manuscripts, in the case of Book 2 on:

> Paris 1910, dated 1471, in the National Library in Paris (Wartelle (1963) no. 1396)

In the absence of a critical apparatus or inspection of Paris 1910, it is impossible to tell what alterations of his source Karsten made, but it seems certain that he made some 'improvements'.[5] In Heiberg's judgement this manuscript is a descendant of:

> K Marcianus 221, fifteenth century, in the library of San Marco, Venice (Wartelle (1963) no. 2122)

Books 2-4 of K were copied from F and corrected by Bessarion on the basis of Moerbeke's Latin translation (K^2). Not surprisingly Heiberg makes very little use of K, but his apparatus includes an extensive, although not complete, record of Karsten's text. As I have suggested, the major gap in his apparatus is the incomplete record of F. However, my main subjective impression from reading the apparatus is the large number of places in which Karsten and F agree against the text printed by Heiberg, frequently producing a *lectio facilior* and, in the case of citations or paraphrases of Aristotle, frequently agreeing with what Moraux prints. (In many cases Moraux's apparatus records variants which agree with what Heiberg prints.) I am not always confident about the distinction between a *lectio difficilis* and a *lectio impossibilis*, but I have fairly often chosen to translate a reading from Heiberg's apparatus rather than what he prints. Such divergences are recorded in the Textual Suggestions on pp. 163-4, and in a note at the relevant place in my text. I do not think of most of these changes as emendations, since there is no point in talking about emendations until a better Greek text of the commentary is available to us.

2. Issues of translation

In English it is customary to distinguish among sun, moon, the stars, and the planets. The usual term for referring to all of them together is 'heavenly body'. The usual Greek word for all of them is *astêr* or *astron*.

I have chosen to translate these two words as 'star' rather than 'heavenly body' because Simplicius often uses that phrase (*ouranion sôma*) and frequently uses it to refer not to what we would think of as a heavenly body, but to the body of the whole heaven or a heavenly sphere. So the reader has to bear in mind that in this translation the general meaning of 'star' is 'sun, moon, star, or planet'.

In Greek astronomy the word 'planet' (*planetês*, usually plural) refers to the moon, sun, and the five known 'planets', Mercury, Venus, Mars, Jupiter, Saturn. 'Planet' occurs relatively rarely in the commentary, usually in connection with other writers or when Simplicius is quoting a passage in which Aristotle uses the word; see the Greek-English Index under *planetês*. In the translation I have used the customary names for the five planets, although, of course, the Greeks associate them with their own divinities, Hermes, Aphrodite, Ares, Zeus, and Kronos. In the commentary on Chapters 1 to 9 Simplicius refers to each of Venus and Mercury in the standard Greek way as the star of Aphrodite and the star of Hermes (454,19-20), but he is also willing to talk about the distance of Aphrodite or Hermes (468,24), the motion of Hermes (415,28), and the spheres of Kronos and Zeus (447,29-30). So it looks as though Simplicius is willing to refer to the five planets simply with the name of the corresponding divinity. Sun and moon are always referred to directly as *hêlios* and *selênê*. I do not capitalise 'sun' and 'moon'; neither does Heiberg, except at 468,24 in a passage in which Simplicius also refers to Venus and Mercury.

The word 'planet' comes from the verb *planasthai*, meaning 'to wander'. Simplicius uses *planasthai* either as an infinitive or as a present participle. As an infinitive it always occurs in an expression like 'star (or sphere) which is said (or thought) to wander', an indication of Simplicius' Platonist belief that the planets don't 'really' wander. When it is used as a participial adjective and the noun it modifies is either explicit or it is reasonably clear what the noun is, I have translated the participle 'planetary' except when the noun is 'star', in which case I have translated 'planetary star' as 'planet'. When the participle is a substantive there are more difficulties. In these cases I have rendered the plural as 'planets'. However, Simplicius often uses the singular to refer to the domain of the planets; here I have rarely (only at 435,3) used 'planetary' and chosen 'the planets' instead.

I have translated the word *aplanês* (unwandering) as 'fixed'. Simplicius most often uses this adjective by itself without a noun. Most freqently it is in the feminine singular and the noun to be supplied is 'sphere' (*sphaira*); so 'fixed sphere' in this translation refers to what we ordinarily call the sphere of the fixed stars. Sometimes 'fixed' is in the plural and the noun to be supplied is 'star'. I list here all the places where *aplanês* modifies a noun:

fixed heavenly body (= the fixed sphere): 408,3

fixed heaven (= the fixed sphere): 420,36; 444,28; 459,26; 487,14
fixed stars: 415,22; 444,18.28; 453,12; 454,25; 455,9; 490,27; 537,14; 549,5 (fixed star at 445,15)
fixed sphere: 449,2; 453,14; 462,13; 514,16; 548,27
At 455,5 the words *aplanê tôn astrôn* (the fixed among the stars) is quoted from Plato

Aristotle does not use the word *aplanês* in *De Caelo*. He normally uses *endedomenos*, which I have translated 'fastened'.

The word *to meson* is sometimes translated 'middle', but it has frequently seemed to me necessary to translate it as 'centre'. Because of this fact I have chosen to translate the word *kentros*, usually translated 'centre' as 'central point'.

A considerable portion of the commentary is very close paraphrase of the text being commented on, so close that the line between paraphrase and quotation is frequently difficult to draw. Heiberg indicates what he takes to be quotations by spacing letters more widely; I use single quotation marks. In general I have identified fewer quotations than Heiberg does, and have tried to restrict quotation marks to strings of at least five or six words which are strictly identical with the corresponding Aristotelian passage. The choice is to some extent arbitrary since Simplicius sometimes omits particles which are grammatically necessary in an Aristotelian sentence but unnecessary when the sentence is embedded in a larger context, and he sometimes merely varies Aristotle's word order or inserts a word which is only implicit in the text he is dealing with.

Notes

1. See especially the unpublished thesis, Hoffmann (1981), e.g., p. 18. I have relied heavily on this work.
2. So Hoffmann (1981), pp. 240, 293.
3. A fact first noticed by Peyron (1810).
4. See Allan (1950) and Bossier (1987), pp. 289-90 and 320-5.
5. See Bergk (1883), p. 143, n. 1 and p. 148.

Simplicius
On Aristotle
On the Heavens 2.1-9

Translation

On the Second Book of *Aristotle's On the Heavens* 365,1

[A summary of Book 1]

As was said previously,[1] the present treatise, following on the discussion of natural and corporeal principles which he has presented in the *Physics*, has as its purpose in the first book the discussion of the simple bodies, the first things constructed from the principles, both the body which moves in a circle and the four sublunary bodies. He demonstrated on the basis of the simple motions that there are five simple bodies and that the body which moves in a circle is neither some one of the four <elements> nor a composite of them <1.2>, but is simple and superior to them in substance (and accordingly in power) and that it does not come to be and is not destroyed, since things that come to be come to be from opposites and things that are destroyed are destroyed into opposites <1.3>, but the body which moves in a circle has no opposite <1.4>. And he proved *per se* and universally that this body is finite because there is no infinite body <1.5-7> and furthermore that it is one <1.8-9> and that there is nothing, neither body nor void, outside the heaven; and at the same time he demonstrated as a consequence of the heaven being of this sort that the cosmos also neither comes to be nor is destroyed and is finite and one, and he proved at the same time the properties which the cosmos has because of the heaven, such as that there is nothing left over outside of it, but that it is a whole composed of the entirety of corporeal structures <1.9, 279a11-b3>. Having demonstrated these things and taking up the discussion of what neither comes to be nor is destroyed, he proved in many ways that the heaven and consequently the whole cosmos does not come to be and is not destroyed, and that it is the case neither that it comes to be but is not destroyed, as some people[2] have been thought to say, nor that it does not come to be but it is destroyed, but rather that it is eternal in each direction. And after that he showed universally that coming to be and being destroyed and not coming to be and not being destroyed follow from one another <1.10-12>.

[A summary of Book 2]

Next, having demonstrated these things on the basis of many consid- 366,1
erations in the first book, he makes the conclusion of them the

starting point of the second book, in which he articulates the remaining issues relating to the heaven:

(i) the fact that motion in a circle and rest in the same <position>[3] belong to the heaven by nature and not because of some necessity; <2.1>
(ii) the issues relating to the spatial dimensions in the heaven and their limits, up and down, right and left, front and back; <2.2>
(iii) the reason why the contrary motion of the planets has arisen; <2.3>
(iv) the fact that the heaven is spherical in shape; <2.4>
(v) the reason why the first heaven moves in the same direction in which it now moves and not in reverse; <2.5>
(vi) the fact that the motion of the first heaven is uniform and not non-uniform; <2.6>
(vii) the issues relating to the substance of the stars themselves <2.7> and to their shape, <2.11> order, <2.10> and motion; <2.8-9>
(viii) the reason why, given that the fixed sphere has a single motion, it is not always true that the things which are more distant from it, such as the sun and moon, have more motions but those in between <the fixed sphere and the sun and moon> have the most motions; <2.12.291b24-292a9; 292a14-b25>
(ix) the reason why each of the spheres which are said to wander contains one star, but the fixed sphere contains as many as it does; <2.12.292a10-14; 292b25-293a12>
(x) and tenth of all, the issues concerning the earth, not in itself, but concerning the relation it has to the heaven, that, having the relation of central point, it lies in the centre of the heaven and is itself also spherical, and the heaven moves around it, while it remains fixed, and it has an extremely small size in comparison with the heaven. <2.13-4>

Bringing things together, the general topics of the book are three in number: they concern the heaven, the stars, and the earth.

[Chapter 1]

283b26-284a2 On the basis of what has been said it is possible to believe that the whole heaven [has not come to be and that it is not possible for it to be destroyed, as some say it is, but it is one and eternal, having no starting point or termination of its whole duration and having and containing infinite time in itself.

283b30 And <it is also possible to believe this> on the basis of the opinion of those who speak differently and generate the heaven; for if it were possible for things to be in the former way, but not possible for them to be in the way described by those who

> say the heaven comes to be, this fact should add great weight] to the belief that the cosmos is immortal and eternal.[4]

Alexander,[5] who considers the entire first book to be about the whole cosmos,[6] says that this conclusion also concerns the cosmos and makes clear that not all the cosmos has come to be, since the body in it which moves in a circle and which is the greatest part of the cosmos, has been proved not to come to be or be destroyed. But perhaps <Aristotle> has not said 'the whole heaven' to make a contrast with some part of the cosmos which has come to be. For, if the sublunary elements come to be and are destroyed but have their coming to be and destruction by changing into one another, they do not come to be in the sense of previously not existing, nor was there a time when either they themselves or the species of things composed from them, animals and plants, were not. But perhaps the words 'the whole heaven' are said to make a contrast with sublunary *individuals*, since, e.g., Socrates came to be when he had not been previously. But such things are not in an unqualified sense parts of the cosmos, but what one might call products; for, if they were parts in the strict sense, the whole cosmos would not be forever at all, since some of its parts would be destroyed. But <if it exists forever>, it is necessary that the entirety of it have its existence in certain things which stand. So, as Alexander says, the words 'the whole heaven' have not been said just about the body which moves in a circle, but about the whole cosmos. And what is said is that the cosmos has not come to be, and that it is not possible for it to be destroyed, since it is not the case that because some bodies change into one another, some parts of the cosmos are destroyed, but, as he said in arguing against the associates of Empedocles and Heraclitus who say that the cosmos comes to be and is destroyed in turn,[7] the cosmos does not come to be and is not destroyed, but it remains eternally and its arrangements interchange with one another. And we would also say the same thing, since what it is to be the cosmos itself is for some things in it to remain forever and for whatever things in it which come to be or are destroyed to change into one another. For perhaps it would be true to say the opposite – that the cosmos would be destroyed, if these things were not in such a way that the cosmos was a cosmos.

Let it be the case that these things are well said by Alexander, provided that just this is added: that all the things which have been said apply primarily to the body which moves in a circle. For the primary discussion in the first book was also about this body; and not coming to be or being destroyed belongs to the whole cosmos because of this body, and it has made the cosmos one and unique. It is also clear that what is said next, beginning with the words 'Therefore it is well to be persuaded'[8] is said about the heavenly body, and the discussion progresses in a unified way with continuity and does not seem to change subject anywhere. And also the words 'having no starting point or termination of its whole duration' (that is, of the temporal extension of its life) <confirm what I am saying>. For this extension is called the duration of each thing, and it is a kind of image of duration in the strict sense[9] insofar as it is a bringing together of a whole life. And so this beginninglessness and endlessness belongs primarily to the body which moves in a circle and through it to the whole cosmos as well.

368,1 To the words 'having no starting point or termination of its whole duration' he <Aristotle> adds with complete precision 'having and containing infinite time in itself'. For if it had come to be at some starting point of time, it is clear that a time would have existed before its existence, and if it were destroyed at some particular time, there would be a time after its existence. So, with the words 'having and containing infinite time in itself' he has indicated that time does not exist before or after it. For time has come to be together with the heaven, as Plato says;[10] it does not exist before or after the heaven, but its infinity makes the temporal extension of the heaven infinite. (One should understand that the the phrase 'infinite time' does not mean '*actually* infinite time' but 'time that comes to be infinitely'. For time is such as to have its existence in coming to be.)

But in what sense does he say that the heaven *has* and *contains* time? Is it, as Alexander thinks, that the heaven is what makes time, since time is the measure of its motion, and there is no time outside the heaven, but what makes somehow contains what is made? Or does what Alexander says deserve our attention? For if time is the measure of motion, what is measured would not be the cause of what measures it, but vice versa. And if time has come to be together with the heaven, how can the motion of the heaven be the cause of time? Perhaps, then, there being many good things which fill out the heaven and the whole cosmos, time is one of these and the most authoritative; and, as Plato says, the demiurge, wishing to make the cosmos more like its paradigm, which is enduring, has made time an image of duration and added it to the cosmos.[11] So it is reasonable that time is a contribution to the perfection of the cosmos, and does not contain it by existing before or after it, as some people think,[12] but time is possessed by the cosmos and partially contained by it, and time exists as filling out the heaven (that is, the whole cosmos).

(283b30) He says that on the basis of what has already been proved it has become believable that the heaven and the whole cosmos do not come to be and are not destroyed, and further on the basis of the refutation of those who think the opposite and say that they do come to be and are destroyed. For, if it is possible that the truth is as we
369,1 say and the cosmos does not come to be and is not destroyed, but it is not possible, as they say, either that it has come to be and is destroyed or that it does not come to be but is destroyed or that it does come to be

but is not destroyed, 'this fact should add great weight to the belief that the cosmos is immortal and eternal'. 'Immortal' is said with respect to the uninterruptedness of life, 'eternal' with respect to the uninterruptedness of substance, as we have also learned in Plato's *Phaedo*.[13]

The precision of Aristotle's words together with the prudence of a philosopher is wonderful. He says that 'it is possible' that the things which he has proved hold in this way. However, the conclusions which he has drawn concerning the fact that the heaven and the whole cosmos do not come to be and are not destroyed are not just possible, they are necessary. And he says, 'but not possible for them to be in the way described by those who say the heaven comes to be'. So then if it is not possible for them to be in the way described, it is necessary that they not be in that way. However, the elimination of a thesis of other people does not lead to our thesis with necessity. That is why he says, 'this fact should add great weight to the belief'; but he does not say 'demonstration'. Why so? Have not the things which have already been proved on this subject been proved precisely? Rather, in the case of things for which it is necessary that they either be in one way or the other, the person who eliminates the possibility of what other people say establishes the alternative which is left. If, then, it is not possible that the cosmos either come to be or not come to be but be destroyed, it is clear that it is necessary for it not to come to be or be destroyed. But, even if something is established by necessity when other things are eliminated, this thing would not thereby be said to be established through constructions and demonstrations appropriate to it, since this method of argumentation is indirect and does not come about in a primary way.

Alexander remarks on the words 'and it is not possible for it to be destroyed' and says they indicate more than that it will not be destroyed. For since, he says, Plato thought that it is destructible, but will not be destroyed because of the will of god,[14] <Aristotle>, having proved that this is impossible, reasonably says 'and it is not possible for it to be destroyed' since he has proved that this is impossible. However, Plato may also have said that it is not possible for it to be destroyed, but this belongs to it because of the will of god, without saying that this belongs to it in a pointless way; rather, even if it is finite in its own nature and has a finite power, nevertheless because it is directly derived from an unmoving cause and reverts to it, it is suitably disposed to receive the uninterrupted motion produced in it by and through that cause.

284a2-14 Therefore it is well to be persuaded that [the ancient 370,1
accounts which are especially a part of our patrimony are true: that there is something immortal and divine included among

the things which have motion, and it has the kind of motion of which there is no limit, but rather it is the limit of other motions;[15] for being a limit is of the things which contain,[16] and this motion, which is complete, surrounds and contains the incomplete motions which have a limit and cessation, whereas it has no starting point or end, it is unceasing throughout infinite time, and it is the cause of the start of some of the other motions and it receives the cessation of the rest. The ancients assigned the heaven and the upper region to the gods on the ground that it alone is immortal.

284a13 The present account bears witness that it is not destroyed and does not come to be] and further is not affected by any of the difficulties to which mortals are subject ...

He has also set out in a very skilful way and before giving demonstrations the generally held beliefs that the heavenly body does not come to be in the first book when he said, 'For every human has a conception of the gods, and everyone who acknowledges gods – both barbarians and Greeks – assigns the upper region to the divine, obviously on the grounds that the immortal[17] is to be associated with the immortal, since it cannot be otherwise.'[18] And now again after the demonstrations he adds the witness making belief more firm based on the ancient and patrimonial accounts. For when he presents the arguments based on generally held opinions before his demonstrations he makes them more briefly and quickly than he does the demonstrations, but he nevertheless creates the impression of a kind of paralogism or inability to give a demonstration; but such arguments, when given after the demonstrations, produce a more secure and deeply felt belief in the demonstrations. And that is why Aristotle makes use of these considerations both before and after his demonstrations and why Plato customarily used myths after his demonstrations.

He says now that it is also well, while remaining confident about the demonstrations, to be persuaded still more that the ancient accounts of earlier people, and especially those concerning the gods which have been handed down from father to children, are true. For not all the ancient accounts are a part of our patrimony, but just those which are handed down from our fathers and countrymen, and concern most of all divine reverence and worship of the gods, because all people who share in these as things handed down from gods strive to preserve them unchanged. Among these beliefs is the view that all heavenly bodies which derive their motion from themselves are both natural and divine, that they have soul, and that they have an uninterrupted motion, 'the kind of motion of which there is no limit' and, being immortal, they are akin to the gods. And so he says that the discussion along with the demonstrations has given witness to these things, proving that the heaven does not come to be and is not destroyed.

What makes especially clear that belief in heavenly divinities is inherent to human souls is those with atheistic preconceptions who slander[19] heavenly things. For even these people say that the heaven is a dwelling place of the divine and his throne[20] and that only it is sufficient to reveal the glory and superiority of god to those who are worthy. And what could be more august? But nevertheless, as if forgetting these things, they consider things which are 'more worthy to be thrown out than dung'[21] to be more honourable than the heaven, and they strive contentiously to dishonour <the heaven> as if fighting against an outrage which had happened to them.[22]

Humans consider the motion of the heaven to be unceasing and uninterrupted, reasoning on the basis of the observations handed down in succession which indicate that the motion is unchanging and always the same; for it would not be possible that anything whatsoever which is destructible remain in sameness for an infinite preceding duration. I also think that the beauty of heavenly things, which is beyond sublunary beautiful things by many measures and which naturally causes souls which have not been previously corrupted to turn back to it, is sufficient for acknowledging that heavenly things are divine; and I also think that their active control of sublunary things and the fact that they contain all bodies are sufficient. And this is a special feature of the divine receptacle: generating all things and containing everything in itself and not isolating any of the things which it creates.

But how can he say that the motion itself of the heaven, which has no limit, 'is the limit of other motions'? For how can one motion be the limit of another, and especially how can a more universal motion be the limit of a more particular one? How can it be that he seems to make a syllogism from two affirmative premisses in the second figure[23] by saying:

(i) A limit contains;
(ii) the heavenly motion, which is complete, contains;
(iii) therefore, the heavenly motion, which is complete, is a limit.

This way of combining premisses in the second figure is unsatisfactory since it implies no necessary conclusion. However, Aristotle has carried out the syllogism in the first figure with the following premisses:

(i) The motion of the heaven, which is complete, contains the other motions;[24]
(ii) that which contains the other motions is a limit of the other motions.

<The second premiss is correct>; for it is a peculiar feature of things which contain to be a limit of that which they contain, as he himself has indicated with the words 'being a limit is of the things which contain'. I do not think that these words are equivalent to 'a limit is among the things which contain', as Alexander says they are; for if a limit were among the things which contain, a limit would be a more particular thing than that which contains, and it would not be true to say that what contains is a limit, just as it is not true to say that an animal is a man since man is among the things which are animals; for more particular things are not predicated of more universal ones, but vice versa.

Alexander says <of Aristotle's words>:

> He does not say this as a syllogism, but as indicating that the complete motion has a certain similarity to a limit, since both a limit and the complete motion contain.

But it is manifest that the motion of the heaven, which is unceasing, contains all the other motions which have a starting and a finishing point, since it exists before them, together with them, and after them; and a certain part of the whole motion, that is, the annual revolution, is a measure of the yearly life of mortal things and of their change of any kind. So, just as both the starting point and the limit of the daily or annual revolution is contained in the whole motion, so too is each of the sublunary motions; and, therefore, the whole motion, which does not have a limit, is also said to contain the other motions and be their limit. For both the starting points and the limits of transitory motions are delimited by the motion which is uninterrupted. For that motion is uninterrupted because anything of it which is taken is not only the limit of the motion which precedes it but also the starting point of the motion which is to come. But how could something in which the same thing is limit and starting point have an interruption or, in general, have a starting point or a finishing point which is determinate?[25] Since, then, the motion of the heaven is of this kind, it is reasonable that it is the cause of the start of some of the other motions and that it receives the cessation of the rest. And it is the cause of starting because all coming to be comes from it; but it is not the cause of cessation in the same way because the motion of the heaven is not in a primary way the cause of destruction, just as the demiurge-god is not, but rather destruction accrues necessarily to things which cannot exist forever and sickness accrues to those who are not constituted so as to be healthy forever. Nevertheless, it does receive the cessation of things that cease; and because, just as the starting point is in it, so too the cessation is also in it, or rather, because the destruction of one thing is the coming to be of another, it also returns the dissolved elements in composites to their own entireties. And it seems to me that in these matters Aristotle is adhering to the Platonic statement which the demiurge-god makes to the heavenly gods in the *Timaeus*: 'Give them nourishment to make them grow, and, when they decline, receive them back again.'[26]

(284a13) So, having said that the ancients assigned the heaven to the gods because of the uninterruptedness of its motion and its immortality, he adds, 'The present account bears witness that it is not destroyed and does not come to be', adding demonstrative precision to the undemonstrated and natural conceptions of the ancients and demonstrating in addition that 'it is not affected by any of the difficulties to which mortals are subject'. <He does not say> that it is not affected without qualification, since perfected heavenly bodies do undergo things by the action of one another, as is made completely clear by the illuminations of the moon produced by the sun and the different actions of the different <astrological> combinations[27] on sublunary things. But 'it is not affected by any of the difficulties to which mortals are subject'. For he proved that it is not subject to alteration in the previous book[28] when he denied that it underwent the passive alterations which go together with things that come to be and are destroyed.

But if, <Alexander> says, heavenly things are not subject to affec- 373,1
tions of this kind, how can it be that the lunar sphere is not heated by the sun but the *hupekkauma*[29] and the air are heated by it, even though what is between them and the sun undergoes nothing? They[30] solve <this difficulty> by saying that not just anything, but only what is <appropriately> constituted, is acted on by something. Accordingly, hearing does not apprehend colours, and eyes do not apprehend sounds because the one is not affected by the other. And they take what happens in the case of the sea creature, the sting ray, as clear evidence that a chance thing does not affect a chance thing and that things in between are not always affected by what passes through them; for when a sting ray is caught in a fishing net, the people holding the ropes of the net are numbed, but the ropes do not suffer anything of this kind because they are not capable of being affected in this way. And they say that those who swim above the electric eel are numbed and sink, although the water in between is presumably not numbed. Therefore, there is nothing anomalous if the body of the lunar sphere transmits the heat of the sun to sublunary things but is not itself heated in a passive way. For a certain kind of coming to be and destruction always go together with an alteration of this kind.

However,[31] nothing prevents the lunar sphere, while being affected by the sun in a vital and perfective way, from transmitting solar rays to the sublunary world and this being affected by them in the way in which it is constituted to be affected. And it is constituted to be heated passively and to be divided;[32] for participations of other stars, which are not hot or cold in the sense of affective heat or coldness, nevertheless also produce such things in the arrangements here. Consequently the moon is illuminated by the sun and it transmits the light to us, but it does not undergo being heated in a passive way nor is it divided by the light, but sublunary things share in the light in a passive way.

Alexander clearly agrees that the things <Aristotle> says here are spoken about the aitherial body which moves in a circle and not about the whole cosmos. Why then has he thought that what was said before these things was said about the whole cosmos only,[33] if there is a single continuity for what is said?

<Aristotle> does not say 'the heaven and the upper region' because he believes that what surrounds[34] and the upper <region> are the same, since the upper and the lower are limits of a straight line, but they are opposite to one another; and the upper region is that towards which bodies having lightness move, just as the lower region is that towards which heavy bodies move. But he calls what surrounds 'upper' either because he is going along with ordinary opinion or because what surrounds lies just above the upper <region>, and it would not be unreasonable to call it upper.

At this point he has finished up the things said in the first book about the heaven in the strict sense and secondarily about the cosmos.

284a14-b5 In addition to these things <the present account bears witness that the heaven> is without effort [because it requires no additional necessary force holding it and preventing it from moving in another way which is natural to it, since that kind of motion would be laborious – and the more laborious the more eternal it was – and foreign to the best arrangement.

284a18 Therefore one should not accept the myth of earlier people who say that the heaven requires some Atlas for its preservation. The people who composed this story seem to have the same conception as later people; acting as if all the upper bodies had weight and were composed of earth, they mythically assigned to the heaven a being having soul constraining it.

One should not accept this way <of thinking> nor the view that it has been preserved for all this time because of the vortex from which it acquires a motion faster than its own inclination <downward>, as Empedocles says.[35]

Nor is it reasonable that the heaven remains fixed eternally because of some soul which constrains it, since it is not possible for this kind of life to be painless and blessed for a soul; for it is necessary that the motion involve force – since it moves the first body in a way different from its natural motion and moves it continuously – and that the soul be without leisure and deprived of all mental freedom, if it does not have the rest which attaches to the soul of mortal creatures, i.e., the relaxation of the body which comes during sleep; rather it would be necessary that it be held eternally and relentlessly by the fate of some Ixion.[36]

So, if it is possible for the first motion to be the way we have described, not only is it more suitable to think this way about its eternity, but also only in this way can we give harmonious accounts which agree with the oracle[37] concerning god.] But let this be enough of such discussions for now.[38]

Having finished up the things demonstrated in the first book – that the heaven does not come to be and is not destroyed and is beyond 'any of the difficulties to which mortals are subject'[39] – he here establishes what remains: that its motion in a circle and its rest in the same <position> are without effort and natural, not forced or by necessity. He is right to combine force with necessity because not everything necessary is forced, since eternal things are necessary, but they are not forced. He proves that the circular motion is without effort and natural as follows:

(i) The heaven requires no additional necessary force (a force which, when it is natural for <the heaven> to move in a different way (e.g., downward), prevents it by restraining its inclination and making it move in a circle);
(ii) that which requires no necessary force for its own motion moves without effort;
(iii) therefore the heaven moves with its circular motion without effort and naturally.

And he proves that what needs no additional necessary force moves without effort from the fact that what needs necessity is 'laborious' and 'foreign to the best arrangement'. For the best arrangement does not involve necessary force, but exists in natural completeness. For it is not only the case that what is forced is laborious, but also what is laborious is forced. So if what is laborious is forced, what does not need necessary force would be without effort.

That the heaven does not need a necessary force is also clear from these two things: (i) being divine, it has the best arrangement, and (ii) what has the best arrangement does not need a necessary force, since what needs a necessary force is foreign to the best arrangement, which is natural and not forced.

(284a18) And it is also clear from the statements of those who, saying nothing precise, make some necessity responsible for this kind of motion or rest of the heaven. For some of them introduce a mythical necessity which causes the heaven not to fall but to keep revolving above, as Homer introduces Atlas when he says:

> Atlas holds up broad heaven with mighty necessity,[40]

and

> He holds the tall pillars which keep earth and heaven apart.[41]

Others make natural necessity responsible for the heaven not falling down, the vortex dominating the proper inclination of the heaven, which is weaker than the vortex – this is said by Empedocles and Anaxagoras.[42] And others say that the necessity involves soul, that
375,1 the soul of the heaven constrains the motion to remain the way it is eternally, as Plato has been thought to say of the soul of the heaven in the *Timaeus*:[43]

> Extended in every direction from the centre to the last heaven, … it began an unceasing life of thought for all time.[44]

But the business about Atlas introduces no demonstrative necessity both because it is a myth and because the people who composed it have themselves seemed to believe that heavenly things have an earthly weight and consequently need some necessity involving soul to hold them up. Accordingly, if it has been proved that heavenly things have neither weight nor lightness, it is clear that the myth would be off the mark, a piece of fiction made up under the impression that heavenly things have weight. If this business about Atlas is some human fabrication put together on the impression that heavenly things have weight, then it has been more than sufficiently refuted because it has been demonstrated that neither weight nor lightness belongs to heavenly things; but if it is really a myth hiding something divine and wise inside itself,[45] let it be said that Atlas is himself one of the Titans around Dionysos, but because he does not act completely wrongly towards Dionysos, that is, because he is active not only in accordance with the Titanic division in the case of the Dionysiac demiurgic creativity, but he also somehow turns away towards the binding power of Zeus, he does his specialities in accordance with both the Titans and Zeus, making distinctions in the case of the greatest parts of the cosmos and holding up the heaven apart from the earth so that the upper does not get confused with the lower; for, as it says in the oracle,

> <Hecate>, as a certain girdling intellectual membrane, distinguishes first and second fire, which strive to mix.[46]

Atlas distinguishes things and holds them up in conjunction with binding them together in one another. For pillars have both the power

to divide and to bind together the things that lie above with those below them.

However, it is also not true that the fact that the circular motion of the heaven and the standing still of the earth in the centre <of the cosmos> endure eternally is due to the speed of the vortex of the aitherial body overcoming the motion of both the heaven itself and of the earth derived from their proper inclination, which is downward, as Empedocles, Anaxagoras, and Democritus are thought to have said.[47] For <they say that> if the aitherial body and the earth had weight, but the circular motion <of the vortex> were faster than their inclination downward and dominated it, the earth, standing in the centre, and the heaven, moving at the perimeter, would remain in the same places, just as, they say, the water in a bowl does not flow out when the bowl is spun around, if the vortex motion is faster than the motion downward of the water.[48] And so <Aristotle> says that it is implausible that it has been preserved for all this time in this unnatural condition; it would be reasonable that <this condition> last a short time with the inclination downward being forced, but it is impossible that it last forever, so that it could not move in this way forever if, having weight, it were of such a nature as to incline downward.

But the third alternative is also unreasonable, the one which says that the divine body remains moving eternally, constrained by its own soul, or that its motion, being constrained by soul, endures eternally. For it is impossible that what results because of necessary force be eternal since this is unnatural, and, since what is unnatural is a falling off from the natural, it is not eternal; for it is necessary for what is natural to be <eternal>, and it is impossible that both <the natural and the constrained> be at the same time. He indicates the resulting impossibility in the case of what is moved when he says that it is not reasonable that the heaven remains fixed eternally because of some soul which constrains it. And what comes next concerns the soul which moves <the heaven>, that 'it is not possible for this kind of life to be painless and blessed' for it, but it is without leisure and deprived of all mental freedom, if the first body naturally moves in one way, e.g., towards the centre, but the soul with force constrains it to move in another way in a circle. For it is not just what is moved in an unnatural and forced motion which labours, but also the thing which causes that motion, and particularly if it were causing that motion continuously and eternally. For what is done once would be endurable even if troublesome, since relief from the trouble would be quick, but what is done continuously and eternally would be unendurable. And a life of this kind would be much worse than the life of souls in mortal creatures, at least if those souls, although they cause their own bodies to walk, which is different from the way they move naturally in accordance with the dominant inclination of the elements they contain, nevertheless rest from this kind of activity during sleep and the other rest periods of animals. However, the souls of heavenly things and these things themselves, being divine, would never have rest from moving unnaturally.

Alexander says that the mental freedom <to which Aristotle refers> is theoretical, because theoretical intellect and wisdom of this kind need freedom most of all. But it is clear that Aristotle is talking about the freedom involved in causing bodies to move and that he adds 'mental' because of what Plato says:

> <Soul> began an unceasing life of thought[49] for all time.[50]

And the same holds for 'that it be held eternally and relentlessly by the fate of some Ixion': for this too seems to me to be uttered in relation to Plato's statement that:

> <Soul>, extended[51] in every direction from the centre to the last heaven … began an unceasing life of thought for all time.

377,1 For it seems to me that Aristotle is picking up on this when he says, 'it would be necessary that it be held eternally and relentlessly by the fate of some Ixion'.[52] The myth concerning Ixion[53] says that Ixion attempted to rape Hera, who shaped a cloud to take the place of herself to encounter him; when he had had intercourse with the cloud, Centaurus[54] was born. When Zeus learned of this from Hera, he bound Ixion to a wheel and caused him to move unceasingly on it. Perhaps the myth means that Ixion attempted to carry out some civic or royal intention; this kind of life is to be associated with Hera. Appearing unworthy, Ixion justly became entangled with a certain material image of Hera, that is, a turbid form of such a leading role (which the cloud indicates, being turbid and more material air). Being entwined with this form, Ixion generated a jumble of rational and irrational activities[55] and was bound by the demiurge-god, who determines what every person deserves, in the wheel of fate and generation from which it is impossible to be set free, according to the Orphic saying,[56] which does not offer relief to those gods to whom Zeus assigns alternating in a circular way and drying out human souls from evil.

It is my opinion that Alexander of Aphrodisias obviously understands the words of Aristotle well on other occasions – and does so better than the other Peripatetics –, but, in the case of the things which Aristotle says concerning Plato, he does not seem to me to bear in mind that Aristotle's counter-arguments are directed at the surface import of Plato's statements. But, contesting Plato frivolously in a way, he tries not only, as Aristotle also does, to refute the surface import of what Plato says, taking simpler <people> into consideration, but he also attacks the ideas of the divine Plato and tries to draw consequences from what Plato says, frequently not even attending to the surface import. Consequently I, setting out the truth, which is dear to god and to Aristotle, will here try to do a careful investigation, putting forward the things which Alexander says are the opinions of Plato about the motion of the soul. I do this because of those who deal with his words in a more superficial way and dare to take from them in a way which is hostile to the views of Plato, which is the same as to say hostile to the opinions of Aristotle and to divine truth.[57]

In the case of these words themselves Alexander says this:

378,1

> In connection with this statement, it would be possible to show that, according to Plato, the souls of gods and so also the soul of the cosmos are in a worse condition than souls in mortal bodies, at least if it is better for any soul to be apart from body, but the soul of the cosmos will never be apart from body, since after its creation <its body> is eternal and moved by soul in unceasing motion, as he says (and this would seem to be a case of causing motion which involves force). For, even if the body of the cosmos is not similar to the bodies of other creatures because they are more subject to being affected, nevertheless it is better for the soul of the cosmos to be on its own than to be with a body. But, in general, the previously mentioned anomaly would follow for those for whom the body which moves in a circle is either one of the four <elements> or composed out of them and further is caused by soul to move in a circle, a motion which is not natural for such a body.

Aristotle himself also thinks that the heaven, which is eternal, has soul, since he says clearly in this book, 'the heaven has soul and a starting point of motion',[58] and subsequently he says again, 'we think about these things as if they were just bodies,[59] monads in a certain order and entirely without soul. But we ought to conceive them as sharing in[60] action and life';[61] and Alexander does agree with these statements. So how can he, to begin with, say in this passage that according to Plato the souls of gods and the soul of the cosmos are assigned a worse condition than the souls in mortal bodies, at least if being apart from a body is better, but they will never be apart from their bodies,[62] which are eternal and always have soul? For these were not only Plato's views but also those of that most genuine of the disciples of Plato, Aristotle, and of that most careful of Aristotle's partisans, Alexander. But how, if it is better for divine souls to be apart from bodies, can it be that, according to Alexander, they never attain what is better since their bodies are eternal?[63] And how, comparing divine souls with those of mortal creatures, can he say that the divine souls are in a worse condition according to Plato, when he himself says that divine souls are always with bodies, at least he does so if he accepts that souls in mortal bodies are actualisations,[64] inseparable from bodies? (And he also completely and implausibly misinterprets Aristotle in such matters.)[65] In fact, it is clear that Aristotle does not censure Plato for saying that the heaven is moved by soul since he too says that the heaven has soul and a starting point of motion in itself, which is obviously its soul; for his discussion was about this.

But when Plato says of the soul, 'extended in every direction from the centre to the last heaven … it began an unceasing and rational life',[66] he takes care that someone hearing about the common extension and interweaving <of cosmic soul and body> does not suppose that the soul's motion is corporeal and a pushing by force, since its body is not of such a nature as to be moved in that way.[67] Therefore
379,1 Plato also avoids <making the soul like an> Ixion, since,[68] he makes clear that he does not think that the body is extended together with the soul or interwoven with it in a corporeal sense when he says that the demiurge framed everything corporeal inside <the world soul>,[69] meaning that the corporeal is established in the soul. And you can learn that Plato does not think that the heaven is moved by the soul in a circle by force, but naturally (even if he does believe that the heaven is composed from the highest form of the four elements)[70] from what is written about the motion of the body before it was given soul:

> <The demiurge> gave it the motion which is appropriate for a body, that one of the seven motions[71] most connected with intellect and wisdom, and so he led it around uniformly, in the same place, and in itself, and he made it move by turning in a circle.[72]

But if he says these things before the body is given soul, it is clear that he is describing the natural suitability of the heavenly body for circular motion. How then is it still possible to accuse Plato of saying that the heaven moves one way naturally and is moved in another by the soul? And above all I believe that Aristotle, who did not disdain to write a summary or abridgment of the *Timaeus*,[73] understood the opinion of Plato in the *Timaeus* on these matters.

However, Alexander says:

> Moreover, the body which moves in a circle will not be moved by force by a different soul, since, if soul is going to cause motion by force, it is necessary for it to be together with a body; for it is impossible for a body to be moved by force by something without contact and resistance, and these things require body. Therefore it would be necessary for the <different> soul to be together with a body through the motion of which it would move the body

> which moves in a circle in an unnatural way. And it is clear that the body of this soul would move in a circle, if it is going to carry the other body around with it and force it (and not in a natural way), since only in this way will it give to the <other> body a uniform and eternal motion; but if it were moved in a circle by the soul in it, it would also be moved continuously and naturally forever. For what is moved by its own soul in this way is moved naturally. But if this is so, that other body would be the <first> body which moves in a circle.

Alexander says these things in these[74] words, but how can he not remark that nothing prevents what is in heaven and is moved in a circle by its own soul from also moving naturally?

I think that what Alexander says next is worth more (or not less) attention than the things he has already said. He writes:

> In connection with these passages of the second book we have inquired what motion the soul gives to the body which moves in a circle if it moves in a circle naturally; the inquiry is necessary and absolutely obligatory to propose; we should consider the solutions <which have been proposed>. The opinion of Julian of Tralles[75] was that the soul was the cause of its rightward,[76] uniform, and ordered motion. But Herminus said that the soul was the cause of its moving to infinity, since no finite body has by its own nature a power of motion *ad infinitum*.

And Alexander, in investigating this, says that not the soul of the heaven but the first mover is the cause of its being moved *ad infinitum*. And if he had carried his discussion just this far, I think he would have spoken well, since, just as motion is derived from soul because of its power of self-motion, so self-identical, self-similar, etc. motion *ad infinitum*[77] has come to the heaven from the unmoving intellectual cause. But, after saying that the first mover is the cause of the motion *ad infinitum*, he adds, diverting the discussion into his own doctrines:

> For its own soul, if it is an actualisation, is rather associated with form and power, so that, according to what was said before, since it is a finite thing, this could not be a power of causing motion to infinity.

However, it has been demonstrated by many well-known men that, although Aristotle says that the soul is an actualisation, he knows of two kinds of actualisation, one inseparable and one separable, and in the *De Anima* he definitely speaks of the rational soul as a separable actualisation.[78]

Since Alexander says nothing against Julian it should be said that if soul is the cause of the rightward motion of the heaven, it is clearly the soul of the fixed sphere. For that is what is moved rightward and the other <spheres> are carried around with it. So then, what motion do the souls of the spheres which are said to wander cause in them? For they are moved rightward along with the fixed sphere. But if their own motion is leftward, it remains to investigate a difficulty: what sort of motion is caused by nature, and what sort by soul – and if soul causes the rightward motion, the motion from the east, of the fixed sphere, what motion will nature cause them to have?

Alexander says:

> We are trying to prove that the nature of the divine body is not different from its soul, but it is like the heaviness of earth and the lightness of fire. For, just as when we say that these things are moved by nature we are saying that they are moved in accordance with these powers, and their motion in accordance with their own powers is not something different from motion in accordance with nature, so too the nature of the divine body would reasonably appear also to be the power of its soul.[79] For what else besides this could its nature be? For soul is more complete than nature, and it is reasonable that the nature of the more complete body, in accordance with which nature and soul the body moves in a circle, also be more complete.[80]

So how do these things harmonise with Aristotle's views, since Aristotle clearly thinks that nature is one thing and soul another; in the second book of the *Physics* he defines nature as a starting point of motion and rest for what it is in,[81] and in the second book of *De Anima* he defines soul as the actualisation of a natural organic body, potentially having life?[82] So if all soul is the actualisation of a natural body, and a natural body is one thing, the actualisation – and most of all the separable actualisation – is another, how could nature and soul be the same? But <Alexander> considers the soul to be an inseparable actualisation, and since nature is inseparable from a natural body and in a way comes to be the same thing as form, he jumbles nature and soul together, thinking that the name of actualisation suits both, so that he also thinks that power and nature and soul are the same. Since Aristotle clearly says that we are disposed to think of heavenly things as just bodies without soul, but we ought to accept that they share in action and life,[83] if bodies without soul are also natural and nature is soul in their case, as <Alexander> says, what would be the use of bringing in action and life requiring soul in addition (for the life requiring soul is the practical life)?

But I think that these things are manifest, and what remains to investigate is nothing less than what motions nature and soul give to the heavenly body.[84] And I think it should be said that nature by itself is not sufficient to cause a complete and natural local motion, since not even the motion of the four elements which is produced by their inclination in relation to the entirety <of the universe> is a complete and natural motion in the sense that the cause of their motion is inherent in them; their motion resembles falling rather than a kind of pulling, since it is brought to fulfillment by a power produced in the particular by the universe. For these elements, since they are only natural things, do not contain in themselves the cause of motion in the strict sense, since nature is not self-moving, but causes motion by being moved by something else. Consequently it cannot be said to be more a mover than a thing moved. And parts which are torn asunder from their wholes are deprived of their source of life in the universe and their complete nature, but they are said to have a nature[85] insofar as when they move they are such as to be again in contact with the universe. And the universe which is natural and has soul is moved in the strict sense by soul and through the intermediary of nature, since nature is a kind of life which is in a bodily substratum and is suitable and naturally conducive to a body's being moved by soul through the intermediary of nature when nature transmits the kinetic power of the soul into a body. I believe these things, and I say that Aristotle does too, since he says that soul is the actualisation of a natural body and since in the sequel when he says, 'for we do not see in any other things that from where their motion starts',[86] and again, 'since we have determined earlier that such powers inhere in things which have a starting point of motion',[87] he clearly says that things which have soul, and not natural things, have a starting point of motion, and Alexander also interprets him in this way. And Plato thought these things even before Aristotle, since in the tenth book of the *Laws*[88] he says that the ninth motion, which, being moved by something else, moves other things, is nature, and it has its existence in being naturally conducive to being moved, not in causing motion.

Consequently, if someone were to ask what local motions nature and soul cause the heaven to have, we would say that the soul through the intermediary of nature makes the heaven move in a circle. This is one and the same motion, but the heaven gets its natural and unforced suitability (derived from form itself) for being moved from nature and its active change of place, towards which it inclined because of its nature, from soul, just as it gets from intellect its always revolving constantly in the same way around the same things and in the same thing. For because of these things which are furnished by intellect, the motion of soul, which is given by nature, is brought together with the body and restored to similarity with intellectual activity. And so that divine man,[89] after asking why the heaven moves in a circle, says because it imitates intellect.

(284a35) But, as Aristotle says, 'let this be enough of such discussions for now'.[90] Having proved that it is not possible for the motion of the heaven to be forced in any of the ways mentioned, but that it is naturally necessary, since it is eternal, he adds that not only does not having its motion forced but natural harmonise with the eternity of the universe but it also harmonises with the oracle which we possess about everything divine, namely that it is necessary for it to be without effort and to spend its life in freedom and blessedness. Only if we say that its activity is not forced but in accordance with its nature can we provide indisputably harmonious accounts. He calls this common conception which we have of the freedom from effort and blessedness of the divine an oracle because it inheres more strongly than what is demonstrated and is most firm and unchangeable. And the oracles based on divine knowledge are of this kind; they come forth beyond all demonstration with an unchangeable believability.

[Chapter 2]

284b6-285a27 Since there are some people, [such as those called Pythagoreans (for this theory is theirs), who say that there is some right and left in the heaven, we should inquire whether this right and left is the way they say it is or rather different – if one ought to assign these principles to the body of the universe.

First of all, if right and left belong <to the heaven>, one ought to accept that the principles prior to them belong to it in a prior way. These things have been distinguished in our work on the motions of animals[91] because they are appropriate to the nature of animals. For all such parts[92] (I mean, for example, right and left) are clearly observed to exist in some animals, and <only> some of them are observed to exist in some animals, but the only such parts in plants are up and down. But if one ought also to assign some of these parts to the heaven, it is reasonable, as we said, that what exists first in animals also exist in it. There are three <pairs> of them, and each of them is a principle. The three I mean are up and down, front and its opposite, and right and left; for it is reasonable that all these dimensions[93] belong to complete[94] bodies. But up is the principle of length, right of breadth, and front of depth. And they are applied in another way to motions, but I call principles or starting points those things from where the motions of those who have them first start; and growth starts from above, locomotion from the right, and perception from what is in front – since I call that towards which perceptions are directed in front.

Consequently one should not look for up and down, right and left, front and back in every body, but only in those which, having soul, have a starting point of motion in themselves; for we do not see in anything without soul that from where their motion starts,[95] since some do not move at all, and some do move, but they do not move in every direction in the same way.[96] For example, fire only moves upward, earth only to the centre. In the case of these latter we speak of up and down and right and left with reference to ourselves, either with reference to what is on our right, as augurers do, or with reference to what is similar to our right, as when we speak of what is on the right side of a statue, or when we speak of what has its position in the opposite way, speaking of right by reference to our left, of left by reference to our right, [and of back by reference to what is in front of us.][97] But we do not see any difference in the things themselves, since if they are turned around we will call their opposites right and left, up and down, and front and back.

285a10 Consequently one might be surprised that the Pythagoreans only said that two of these, right and left, are principles, but they left out the other four, which are no less important, since in all animals up differs no less from down and front from back than right from left. Some of these things differ in power only, others also differ in their shapes, and up and down exists equally in all things having soul, in both animals and plants, but right and left does not inhere in plants.

285a19 Moreover, since length is prior to breadth, then if up is the principle of length, right of breadth, and the principle of what is prior is prior, up will be prior to right in coming to be (since priority has several senses).

285a22 In addition, if up is that from where (*hothen*) the motion <comes>, right that from which (*aph' hou*)[98] it comes, forward that to which it goes, in this way too up will acquire a certain power as a principle relative to the others.

285a26 It is then right to censure the Pythagoreans for leaving out more important principles and for thinking that] the principles exist equally in all things.

He has said that the whole heaven is simple, that it transcends the four elements, that it moves in a circle, that it does not come to be and is not destroyed, that it does not increase in size or alter and 'is not affected by any of the difficulties to which mortals are subject', and that it is free from effort since its motion is natural and not forced. He has also reminded us of the common conceptions about the heaven, which are like oracles about god, since humans first called the heaven and the things in it gods because they run and since humans moved from these things to a recognition of the demiurge and the other divine orders.[99] And, having said these things, it is reasonable that he proposes next to give instruction on the parts of the heaven.

There are two kinds of parts <of the heaven>, *per se* parts, such as the spheres and the stars, and relative or relational ones, such as right and left, front and back, and up and down. And these latter are also studied in two ways, with respect to position and with respect to motion. And he first discusses the relational parts since for most people it has seemed disputable that the heaven, which is simple and homoiomerous, has these parts at all, and the Pythagoreans placed right and left in their two columns of ten pairs of things,[100] which they said are the common principles of everything (and so of the heaven). He says that, since in general one should observe these principles relating to right and left in the body of the heaven, which is homoiomerous in this way – i.e. in the body of the universe delimited by the heaven –, we should inquire whether these things are the way the Pythagoreans say or rather different in the way that we think. For since there are three dimensions of body, length, breadth, and depth and two limits relating to each dimension, there are three different pairs in the bodies of animals. For up and down are the limits of length, right and left of breadth, front and back of depth, and 'up is the principle of length, right of breadth, and front of depth'. But these things are the starting points of motions, since the motions of those who have them, that is, of animals, first start from them, 'growth from above, locomotion from the right, and perception from what is in front', and most of all perception which is directed forward. Consequently we are not looking for these principles in all bodies, but only in those which contain in themselves a starting point of motion in both directions in all dimensions, and these are bodies having soul. For even if something without soul, such as a clod of earth or a spark, has a starting point of motion it does not have a starting point of motion involving soul, but a starting point of natural motion. The former proceeds in all directions, but the latter is limited to one; for a clod of earth moves only towards the centre and a spark only upward. And if we sometimes speak of up and down, right and left, and front and back in the case of the parts in things without soul, we speak with reference to ourselves; we speak of right either 'with reference to what is on our right, as augurers do' (for we say that a bird which is on our right is a bird from the right and we call it an omen on the right side in the same way), or 'with reference to what is similar to our right' (as when we speak of the right or left hand of a statue), 'when we speak of what has its position in the opposite way' as with what is thought to be facing us (as in the case of mirrors where right is what is on our left and left is what on our right). In the case of things without soul we do not see any natural difference, since, if the things which are taken with reference to ourselves are turned around, we will call the opposites of what preceded right and left, up and down, front and back. Aristotle has taught us clearly about most of these things in *On the Progression of Animals*, and he now refers to this treatise for discussion. There he writes as follows:

> All living things have a part which is up and a part which is down, since there is not only up and down in animals but also in plants.[101]

And proceeding a little further:

> Front and back belong to things which not only live but are also animals, since all animals have perception.[102]

And again proceeding a little further:

> In those animals which not only share in perception but can change place on their own, left and right are distinguished in addition to the things already mentioned, and[103] each of the two is distinguished in the same way as those others, that is, by reference to function, not to position.[104]

But some animals, namely those such as zoophytes,[105] which do not move locally, will not possess left and right.

(285a10) This being the way things are, Aristotle censures the Pythagoreans on these three points:

First because, although there are six distinctions, they used only two, right and left, and omitted the other four, up and down and front and back.

Second because they have left out the more important distinctions, up and down; and he proves this by means of several arguments. First that either there are no fewer clear distinctions between up and down than between right and left or in truth there are still more and more natural ones in their case than in that of right and left. For example, right and left hands or feet, which do not differ from one another in shape, differ only in power, the right hand being more powerful than the left, whereas the left foot is more suitable for standing still, the right for starting to move; and similarly the left shoulder is more suitable for bearing burdens, the right for motion, but nevertheless these things do not differ in shape; but head is distinguished from feet and branches from roots and in general up from down and front from back not just in power but also in shape. In this way then he censures 385,1
the Pythagoreans for using only two of the six principles and leaving out those which differ in more ways and are more important. And second he censures them because they leave out the more common and universal principles since 'up and down exists equally in all things having soul, in both animals and plants, but right and left does not inhere in plants'.

(285a19) And third he censures them because they leave out things which are prior in nature and use posterior things. For length is prior to breadth, since line is prior in nature to surface and has more the character of a principle than it – point and line are related in the same way –, and since growth in length <i.e. height> in animals is prior to growth in breadth, and since if line is destroyed, so is surface, but line is not destroyed if surface is.[106] So, if length is prior in nature to breadth and the principles of things prior in nature are prior, then the principles of length, i.e. up and down, will be prior in nature to the principles of breadth, right and left. But priority has several senses:[107] there is priority in position,[108] as in people seated at a banquet; in order,[109] as prologues are prior to stories; in power[110] or honour, as a ruler is prior to the people he rules; or in nature,[111] as when the destruction of one thing brings the destruction of another, but not vice versa; in time[112] and coming to be, as a father is prior to his son. And he says that up will be prior to right in coming to be because, as we said, in animals growth in length is prior to growth in breadth. So if length is prior to breadth, but up is the starting point of length, right of breadth, the starting point of what is prior will be prior in coming to be.

(285a22) <Aristotle's> fourth objection in addition to these is that if up is the starting point of the motion which is characterised as growth, right that from which the motion starts, and forward that to which it proceeds, and the motion of growth is more important and substantial for an animal than local motion, it is clear that up will possess a more important power as a principle than the other differences. So, their using right and left and leaving out up and down is absurd.

In this way then Aristotle uses several arguments to argue that the Pythagoreans have left out the more important and fundamental principles. But it is now possible to prove from the differences of the prepositional phrases 'from where', 'from which', and 'to which' that up does not differ from right and front in priority; rather it is just different from them. For if the phrases are different, the starting points referred to by means of them are different.[113]

(285a26) His fifth[114] objection to the Pythagoreans – or second, if one takes all of the first things as one argument saying that they have passed over the more important things – ...[115] is that they considered
386,1 these principles, that is, the principle of right and the principle of left, to exist generally in all things although they do not exist naturally in things without soul, since even if they are also spoken of as in these things, they are so spoken of with reference to ourselves, as was said previously.

<Aristotle> infers that they say right and left are in all things from their taking the principle of left and right as one item from the two columns of ten pairs which they say are the principles common to all things (since they did not study these things just in application to the heaven). What purpose do the Pythagoreans have then, and why, when they speak well, does Aristotle raise objections to the surface

import of what they say? The Pythagoreans placed all antitheses in two columns, one better and one worse – or one column of good and one of bad; <taking> ten symbolically as equivalent to all number, they filled out each column; and they took each of their ten antitheses to indicate simultaneously all the antitheses akin to it. And of the spatial relations they took right and left, because these things indicate at the same time good and bad (at least we speak of a 'right' nature or a 'right' fortune,[116] meaning a good nature or fortune, and we mean the opposite by left), and at the same time because up and down and front and back exist in the things in which right and left do (as Aristotle has determined) but not conversely.[117] It was then reasonable for them to use right and left to indicate the other spatial antitheses. For they called right, up, and front good, left, down, and back bad, as Aristotle himself recounts in his collection of Pythagorean views.[118]

We have, then, indicated the whole purpose of the text before us. As for details, when he says, 'such[119] parts (I mean, for example, right and left)', 'front and back and up and down' are missing from the text; he has left them out because they will be familiar from what he has said.

And when he says, 'but[120] it is reasonable that these[121] dimensions belong to complete bodies', he does not mean length, breadth, and depth, since these belong to all bodies, but the three antitheses, up/down, front/back, right/left. He calls bodies complete if they have not only a life of nutrition and growth like plants or just perception in addition to these things, as is the case with zoophytes, but also desire and locomotion, like the complete animals. That is why he adds, 'but up is the principle of length', and so on.

One should note on the basis both of the text which says, 'for we do not see in anything without soul that from where their motion starts', and also of the interpretation of Alexander (who says: 'by the starting point of motion he might mean their kinetic power, since things without soul also have a certain starting point of motion, at least if they are natural bodies, but they do not have a principle which causes motion nor a kinetic power in themselves, since what moves them is external; but they do have a certain passive principle and power of being moved') the way things without soul have a nature to move, not on their own, but by one thing being moved by another, as he proved in the *Physics*.[122] Consequently, if someone asks how nature moves the heaven and how soul does so, one should not say, as Alexander did, that nature and soul are the same in that case.[123] For how can they be the same if nature is a passive power of being moved which is present in a substratum which is moved, but soul is what causes motion from outside?[124] Therefore, one should not say that soul and nature are the same in the case of the heaven, although both do play a role in the same motion, soul as an external cause of motion, nature as an inherent principle of being moved.

When he says of things without soul, 'some do not move at all, and some do move, but they do not move in every direction in the same way', he refers to things which have already occupied their own place in which they remain thereafter as not moving; and the things which do not move in every direction are things which do not move in all six dimensions, but only up, as in the case of fire, or down, as in the case of earth.

However, we should proceed to what comes next.

> **285a27-31** We have determined previously [that such powers inhere in things which have a starting point of motion; but the heaven has soul and a starting point of motion; therefore, it is clear that it also possesses up and down] and right and left.

Having censured the Pythagoreans for having left out the more important principles of the dimensions and taken as universal principles which are not universal, he next explains what remains, namely how one should understand these antitheses relating to the dimensions in the case of the heaven. And first he proves that one should also look for these principles in the heaven in the following way:

(i) The heaven has soul;
(ii) what has soul has in it a starting point of motion which causes it to move;
(iii) the powers themselves <of moving in these ways> belong to a
388,1 thing which has a starting point of motion of this kind (since it has been determined previously[125] that one should not look for up and down, left and right, front and back in every body, but only in those which, having soul, have in themselves a starting point of motion which causes them to move);
(iv) therefore the heaven has these powers.

For even if the heaven does not grow, there is nothing to prevent up and down belonging to it, since it is also not the case that if something does grow and it ceases growing, up and down straightaway cease to belong to it. But when he proved in the preceding[126] that in the case of things that grow growth in length is upward, he used this as a clear sign that up and down belong to animals and are different from right and left. And this being so, he <now> is proving universally that up and down and front and back belong to those animals which are complete and in which there is right and left and which have in themselves the starting point of causing locomotion. For growth is prior, and then comes perception, and then desire and the local motion which is produced by desire.

Since some of the philosophers whom I worship think that Aristotle says that the heaven has soul in the sense that it has life inherent in its body in terms of which it has a starting point of motion, but not in the sense that it has a rational soul, I ask them first to accept that Aristotle says that what has soul is one thing, natural things another, and then to determine what having soul in the case of the heaven is on the basis of those words in this very book in which he says about heavenly things:

> We think[127] about these things as if they were just bodies,[128] monads in a certain order and entirely without soul. But we ought to conceive them as sharing in action and life.[129]

For also, according to <Aristotle> practical activity attaches to the rational soul.[130]

> **285a31-b8** There is no need to raise a difficulty [about how there can be right and left in the universe <just> because the shape of the universe is spherical, all its parts are similar, and they move for all time. Rather one should think that the situation is as if someone were to place a sphere around something which also[131] possesses the difference between right and left in shape; it would then have a differentiating power, but would seem not to because of the uniformity of its shape.
>
> 285b5 One should think in the same way about the starting point of motion, since even if it never started to move, it would nevertheless necessarily have a starting point from where it would have started to move, if it had started,] and from where it would move again, if it were to stop.

Having demonstrated that one ought to look for such principles in the heaven, he sets out two objections to be brought against the present argument. And he dissolves one, which is common to all of these antitheses and is based on the uniformity of the parts of the heaven: how can the heaven have right and left, up and down, front and back, if it is spherical and homoiomerous and these <dimensions> ought to differ in terms of shapes? Of these he mentions only right and left, but I think he uses them to refer to the others, because animals which have a right and left also have the other antitheses in a still prior way. 389,1
He hints briefly at the other objection, which is to right and left, with the words 'and they move for all time'. For if right in animals is that from which locomotion starts, how, in the case of what is always moving and never starts to move, can one speak of a starting point of locomotion? Having introduced the first objection with several words

and both objections briefly with the words 'all its parts are similar,[132] and they move for all time', he dissolves the first objection on the grounds that one should not determine difference by shape but by power. For if someone were to place a spherical figure around an animal in which the difference is thought to be a matter of both shape and power, he would not eliminate the difference of powers by means of the uniformity of shape.

But perhaps someone will say that if the spherical figure is placed around <an animal> and the difference in shape <of the animal> remains, it would not be surprising if the difference in power also remains along with it; and Aristotle does seem to speak about the placing of the sphere as if the difference in shape remained inside. But perhaps one should not understand the placing of the sphere in this way. For what would be the point of covering with a sphere from outside as if with a cloak? But if someone placed the sphere around to replace the anhomoiomerous differentiated structure and the differences in power remained, the difference would be preserved and the activities of growth, perception, and motion would start from where they start now. In general, in things which are right or left it is not difference of shapes which make the difference; rather it is powers.[133]

(285b5) He dissolves the second objection by hypothesising that the heaven starts its motion at some time, stopping it for the sake of the argument, and investigating from where it would start to move. For in seeking what follows in fact we hypothesise many impossibilities for argument's sake, as when we separate qualities and quantities of material substance although quality and quantity cannot exist apart from substance or substance apart from them. And Aristotle here proceeds in the manner of Plato, who says in the tenth book of the *Laws*:

> If everything[134] were to stop and all things came together … which of the motions of which we have spoken would have to come to be first?[135]

As far as the text is concerned, one should insert a comma before 'but would seem not to because of the uniformity of its shape'.

390,1 **285b8-27** I call the length of the heaven [the distance between the poles, one pole being up, the other down. For we see a difference only in these hemispheres because their poles do not move.[136]

285b11 And at the same time we are accustomed to call, not what is up and down, but what is beside the poles, the sides of the cosmos on the grounds that <the distance between the

poles> is the length; for what is to the side is what is beside what is up or down.

285b14 Of the poles the one which is observed above us is down, the invisible one up. For we call the right <side> of anything that from where the starting point of its locomotion <comes>, and the starting point of the revolution of the heaven is that from where the stars rise, so that this would be right, and where they set would be left. So if the heaven starts from the right and revolves to the right,[137] the invisible pole must be up; for if it were the visible pole, the motion <of the heaven> would be to the left, which we do not believe. It is clear then that the invisible pole is up, and the people who live there are in the upper hemisphere[138] and towards the right, whereas we are in the lower <hemisphere> and towards the left, which is the opposite of what the Pythagoreans say, since they make us up and in the right part and those there <in the other hemisphere> down and in the left part;] but the opposite is the case.

Having proved that one should look for these dimensions in the heaven and having dissolved the objections against this, he next explains how one should understand the dimensions in the case of the heaven. He first finds length from the size of the dimension. For the greatest distance in any body is said to be its length; but the distance along a diameter is the greatest distance in a sphere (since the diameter is the greatest straight line in a sphere); and, although it is possible to take infinitely many diameters in a sphere, one should take a determinate one, and that on the axis is determinate; so this will be the length of the heaven. But the limits of length are up and down, just as right and left are the limits of breadth; therefore one of the poles is up, the other down. But which is which is still unclear. He proves that one should locate the difference of up and down in the poles in one way from the fact that, of the infinitely many hemispheres <which can be> taken, the only determinate ones are those in which the poles are, that is, the northern and southern hemispheres, which are determined by the equator. For all the other hemispheres, e.g. those determined by the meridian or the horizon, are distinct, since meridians and horizons are different and distinct for different locations; unlike the equator, they are not *per se*, but relative to locations. But the hemispheres determined by the poles always move in the same fixed place, since they move around their poles as around fixed central points. Therefore, Aristotle also says that the fact that 'their poles do not move' is a reason for the distinctiveness of the hemispheres having them <as poles>; for since the poles never move and things which move around them always maintain the same distance from them, it is necessary for what moves around them always to occupy the same place.

(285b11) He also proves the same thing on the basis of the common conception. For we are accustomed to call, not the poles, but what is beside the poles sides of the cosmos. So if what is beside each of up and down are sides, but we say that what is beside each of the poles are sides, it is clear that the poles are up and down. He has produced a little unclarity in the text by saying 'not what is up and down' instead of 'not the poles'; for he appropriately adds on to 'not what is up and down' the words 'what is beside the poles'.

(285b14) Having proved that the poles are up and down, he next determines which pole is up and which pole is down. And he explains this in a way different from the supposition of ordinary people when he says that the south pole, which is always invisible to us, is up in the heaven. The scientific person proceeds in this way; he always looks to the truth, and if the opinion of ordinary people harmonises with it, he introduces this after the demonstrations, and it makes the necessity of the demonstration plausible and irresistible for students; but if the opinion of ordinary people is out of tune with the truth he disdains it as if it did not even exist, as he does now. For ordinary people, judging on the basis of relation to us, say that the pole which is always visible to us and seen on high is up and the always invisible pole is down, but he proclaims the opposite of what ordinary people think because he is investigating what is up and down by nature. For, even if some particular animal buried its head and raised its feet up high, its feet would not be up and its head down by nature; and plants, for whom up is by nature around the roots because they take in nutriment and grow there first, are thought to have their branches up relative to us.

Alexander also chastises Aratus of Soli for saying in agreement with ordinary people that the north pole is up when, discussing the axis, he writes:

> And two poles limit it in both directions, but one is not visible;
> the other lies opposite in the north above Ocean.[139]

Alexander objects only to the phrase 'but one is not visible'; he says that Aratus does not indicate <the true> up and down with the word 'visible', but instead refers to up with the words 'above Ocean'. But one should not censure Aratus; for since he announces that he is writing about appearances and he entitles his treatise thus, it is reasonable for him to determine being above or up for the pole on the basis of a relation to us.

<Aristotle> proves that the south pole occupies what is by nature up in the heaven, the north pole what is down (even though ordinary people think the opposite) from the fact that the east is right, the west left. Since this is so, it is clear that the upper pole is the south pole. And he proves that the east is right from the fact that in the case of

particular animals we call right that from where the starting point of their locomotion <comes>, but the starting point of the revolution of the heaven is from the rising of the stars, so that the left is where they set. So, if the motion starts from the right and revolves to the right (since this is better than revolving to the left), and moving to the right is moving towards the front, it is necessary that the south pole be up. It is as if one were to think of a person stretched out on his back along the axis with his head towards the south pole[140] and being carried around with the heaven; his motion would be from the right and to the right; but if he had his head towards the north pole and had his back on the axis, looking towards the visible, his motion would not be from the right nor to the right but opposite to that; and if one were to imagine him turned over on his front, he would have his head towards the north pole and his motion would seem to be from the right, but not to the right; for motion to the right is always towards the front. So Aristotle has been completely precise in adding the words 'and revolves to the right'. For motion from the east is thought by some people to be from the right of the person lying on his front, but it would not be to the right, since it is towards the back. The most wise Homer also gives witness to the fact that the right side of the universe is the east and the left the west when he says;

> whether they are to the right towards dawn and the sun,
> or whether they are to the left towards misty darkness.[141]

Finishing up his discussion, Aristotle says, 'it is clear then that the invisible pole is up, and the people who live there are in the upper hemisphere'.

The things Aristotle has said up to now are consequences of what he has proved already, but what comes next is thought to face many objections. For (i) how can those in the upper hemisphere also be towards the right and those in the lower also towards the left, since up and down are equally distant from right and left? And (ii) how can he say that the Pythagoreans make us up and in the right[142] and those there <in the other hemisphere> down and in the left, since, as he himself recounts in the second book of his collection of Pythagorean doctrines, they say part of the whole heaven is up, part down, and that the down part of the heaven is right, the up left, and that we are in the down? But perhaps (ad (i)) here he does not use the words 'up and towards the right' in accordance with his own view but in accordance with the Pythagoreans; for they coordinate up and front with right, down and back with left. (ad (ii)) Alexander thinks that what is said in the collection of Pythagorean doctrines has been altered by someone and ought to say, 'what is up in the heaven is right, what is down left, and we live in the upper <part>', and not 'in the lower <part>', as is written in the text, since in this way it will harmonise with what is said here, namely that in saying that we live in the lower <part> and consequently in the left (since down has been coordinated with left), we say the opposite of what the Pythagoreans say, namely <that we live> up and in the right. And perhaps this emendation is correct since Aristotle knows that they coordinate up with right and down with left.[143]

Alexander proves that the poles are the up and down of the heaven in the following way. He assumes that the heaven, since it exists as an animal, ought to have dimensional differences and have them to a
393,1 greater degree to the extent that its parts differ to a greater degree, and that the poles differ from each other most of all since they maintain a stable distance from one another in their locations. Assuming these things, which have been previously proved by Aristotle, he makes the following syllogism by division:[144]

> (i) Since the poles ought to differ from each other most of all, it is necessary that they differ with respect to one of the three dimensions in animals, that is, with respect to right and left or front and back or up and down;
> (ii) but to say that they differ with respect to right and left is absurd, since in everything right is that from which locomotion starts, but the poles are motionless (because the whole heaven revolves in the same <position> because of them);
> (iii) but it is not possible to say that they differ with respect to front and back either, because in the case of things which move by nature, that towards which they move is front, but motion for the cosmos is not towards either pole;
> (iv) if, then, the poles do not differ from one another either with respect to right and left or with respect to front and back, but they differ more than the other parts <of the cosmos>, up and down would be the heavenly distance related to them.

Alexander also argues in the following way:

> (i) Up and down are something alongside of that from where the starting point of locomotion comes and its opposite <the end of local motion>; that is, up and down are alongside of right and left;
> (ii) but in a divine body which is spherical and moves in a circle only the poles remain as something alongside of left and right;
> (iii) therefore, up and down are to be found in the poles.

That for animals which move naturally the starting point of local motion is from the right is clear from the fact that those who are about to run put their left foot forward in such a way that they start their motion from their right foot and around their left, which remains fixed. And this putting forward of the left foot is a preparation for the starting of forward motion from what is right. We also place burdens lightly[145] on the left shoulder, leaving the right side free to start the motion. And it would be reasonable that natural locomotion start from what is stronger; but the right is naturally stronger, just as it is also hotter, since hot is in the same column as right, cold in the same as left; consequently right is by nature stronger, and when some one is also strong in the left, we call him ambidextrous.[146] But if right and left relate to the side, but sides are alongside of up and down, then up and down will be in those things in which there is right and left; but right and left will be in those things which move locally on their own, since right is that from where the starting point of locomotion <comes>. But the divine body moves in a circle on its own, and that is local motion; therefore, the divine body has right and left, and, if it does, it also has up and down.

> **285b28-286a2** But with respect to the secondary revolution, [that is, the revolution of the planets, we are up and right and those <in the southern hemisphere> are down and left. For the starting point of motion is the reverse in their case because their motions are opposite, so that it follows that we are in the direction of its starting point, they in that of its finishing point. Let this much be said about the dimensions of the parts <of the cosmos> and their spatial] determinants. 394,1

He has proved that the up of the heaven is related to the south pole, which is always invisible to us, and the down to the north pole and that consequently those in the southern hemisphere live in what is up and right – since, according to the Pythagoreans, what is up is right[147] – but that we live in what is down and left, inferring what is up and down from what is right and left, and taking the east to be the right of the heaven because the starting point of its motion is there. He now reasonably remarks that these things have been said correctly about the fixed sphere, which moves from the east, but that the opposite results in the case of the sphere (taken as one sphere) of the planets. For if in the case of that sphere too right is that from where the starting point of motion <comes>, right for the planetary sphere would be in the direction of the setting of the fixed sphere. But if this is so, the north pole would be up, the south pole down and 'we are up and right' – as the Pythagoreans may have said[148] – and those in the southern hemisphere 'are down and left'. And the reason is that the starting point of motion, i.e. the right, is the reverse in the case of the planets. And the reason for this is that their motions are in reverse. And he will investigate the reason for this shortly.[149] He says that their motions are 'opposite' instead of 'reverse' since he has proved previously[150] through several considerations and will say again[151] that there is no motion opposite to motion in a circle. But if up and right are starting points, one of growth, the other of locomotion, and down and left are finishing points of them, it follows that in relation to the planetary sphere[152] we live in the starting point and they live in the finishing point. Consequently, ordinary people who say that the north pole is up and we live in the direction of what is up, but the south pole is down and those people live in the direction of what is down, speak correctly – if they are speaking while focusing on the planets.

Someone might reasonably inquire why, although of these dimensions <Aristotle> has given an account of right and left and up and down, he has passed over front and back. It is possible to say that front and back have also become easy to understand on the basis of what has been said; for if a motion which starts from the right goes forward, it is clear that what is above the earth is always forward and what is under it is backward, and if you imagine that person lying on his back with his head inclined towards the south pole, he would always see[153] what is above earth in front of him. Aristotle himself,
395,1 after the discussion of the shape of the heaven,[154] will look for the reason why it moves in the direction that it does,[155] and will then, having found the appropriate place for treating the issue, discuss front, as we will learn then.[156]

285b33-286a2 Let this much be said about the dimensions of the parts <of the cosmos> and their spatial determinants.[157]

Next, finishing up, he refers to what has been discussed as 'the dimensions of the parts' – presumably length, breadth, and depth – and their spatial determinants, up and down, right and left, front and back – for these are the differentiae of locations. But perhaps both phrases refer to the six spatial dimensions themselves. This is the interpretation of Alexander, who says:

> The dimensions of the parts are up and down and the coordinates of these, since the parts stand at a distance with respect to these differences. And <Aristotle> has reasonably added 'their spatial determinants', since up and down and their coordinates are determined by locations. Furthermore, he has determined them spatially and has proved that we live in the lower <part of the cosmos> and those in the direction of the south pole live in the upper.

[Chapter 3]

286a3-8 Since a motion in a circle is not opposite to a motion in a circle, [we should inquire why there is a plurality of motions, although we have to try to make the inquiry from far away, not

> far away spatially but much more because we perceive very few of their features. But let us discuss <this subject> in any case. The reason for these things should be] grasped from here.[158]

According to the order of the issues[159] there would follow next – after the relational difference of the parts of the divine body which exists in both the fixed sphere and in the sphere which is said to wander – the consideration of the difference in their motions: why is there a reverse motion <of the planets>? In fact, in discussing the issue just finished up he reasonably called attention to this reverse motion in proving the difference between up and down in the case of each motion,[160] since there is also a difference between right and left from which the difference between up and down was determined.[161] There is a difference[162] between right and left because the starting points of motion <for the fixed stars and the planets> are different, and these differ because the motions are reverse. Since, then, that difference has been reduced to this cause, it is necessary for the lover of learning to investigate the reason for this, that is, for what reason this contrary motion has come about.

And first he shows that the inquiry is necessary, and then that discovering the answer is difficult. He proves the necessity as follows. Since the fixed sphere moves in a circle, if, he says, there were a circular motion opposite to motion in a circle as motion down in a straight line is opposite to motion up in a straight line, it would be sufficient for the inquirer to answer the person asking why the planets have a motion which is reverse to that of the fixed sphere by saying that, since the circular motion of the fixed sphere has an opposite, it would be absolutely necessary that, since this motion exists, the motion opposite to it also exist. For if one of a pair of 396,1
opposites is natural, it is necessary that the other be so too, since opposites are relative to one another, and in the case of relatives it is necessary that one be if the other is.

But since it has been demonstrated that there is no motion opposite to motion in a circle,[163] an explanation of the contrary motion based on this would be absolutely fruitless. It is necessary to seek another explanation. And, as the noble Alexander says, here acting like a man,[164] it is not possible to make either natural or material necessity responsible for these things, since both <spheres> have the same matter, but it is necessary to give an account of the difference in terms of some divine governance and ordering.

<Aristotle> indicates the difficulty of giving an explanation in such cases when he speaks about making the inquiry from far away; he does not mean that we are at a great distance from them spatially but much more that we are at a distance from them because we perceive few of their features. For if demonstrations are based on *per se* attributes – which he here calls 'features'[165] – and it is necessary for

demonstrations concerning perceptibles to take their starting points from perception, then in those cases in which we are unable to perceive, it will not be easy to furnish demonstrations either. Rather, as Plato says, it is necessary that discussion of such things be <only> reasonable[166] (and Plato says this universally about all perceptibles and images).

I think that <Aristotle> speaks entirely accurately when he says that we are more distant <from the heaven> in knowledge than we are spatially. For even if our distance from the heaven is many times as great as our distance from some visible thing, e.g. a particular tree, nevertheless the two distances are certainly in some ratio, but our ignorance of the attributes of heavenly things stands in no ratio to our knowledge of the way the tree appears. And this also happens in our world, since we do not perceive things from a distance of ten stades at all (unless they are in the sky), but nevertheless their distance stands in some ratio to the distance of what is seen by us.[167]

> **286a8-12** But[168] each thing having a function [is for the sake of that function. The activity[169] of a god is immortality, that is eternal life. Consequently it is necessary that eternal motion belong to a god. But since the heaven is of this kind, because it is a certain divine body, it consequently has a circular body] which moves in a circle naturally forever.

Having proposed to state the cause of the contrary motion of the heavenly spheres, he first takes as an axiom that each thing having a function is for the sake of that function. He calls the product and activity <of a thing its> function, as he makes clear by adding that 'the activity of god is immortality'. For the capacities involved in arts and sciences are also for the sake of their activities, as can be seen if each of the theoretical, productive, and practical <arts and sciences> is examined. For knowledge of the attributes of magnitudes is the end of geometry, building a house of housebuilding, playing a flute of flute
397,1 playing. And even more so in the case of natural things: for a horse is for the sake of running and a human being for the sake of practical activity and contemplation. So too each of the parts of an animal's body is for the sake of its activity, which the animal needs.

Since, then, each thing of which there is an activity is because of that activity, but the activity of the divine body (which he now calls god) is immortality, the divine body is immortal and has its being in immortality. It is then true to say that the heaven is a god;[170] and a god of this kind is a divine body; and the divine body has an eternal life; and what has eternal life has an eternal locomotion; and what has an eternal local motion moves in a circle, since this has been proved to be the only motion capable of being eternal because only it is continuous;[171] for there is no infinite motion in a straight line, and finite motions are divided by a rest period when they turn back. So, he has proved that the heaven moves in a circle from the fact that, being a god and a divine animal, it has a locomotion which is eternal, and only a circular motion is of this kind.

We should consider each of his premisses, and first the first axiom of the argument.[172] For (i) if substance is everywhere stronger and produces activity, but in the case of arts and sciences positive conditions[173] produce activities, how can it be true that each thing having an activity is for the sake of that activity? Secondly, (ii) how can immortality be the activity of the divine animal, since immortality is a component of its substance, as mortality is of a human being – and it is clear that mortality is a differentia and affection of a human being, not an activity? (Ad (ii)) For the present we should say one thing concerning this, namely that since gods are not subject to affection, as has been proved previously,[174] the participations which complete <their substance> are not passive, as is the case with a human being, but active. So, just as the rational and intellectual element in them is not passive, but active, since they do not possess anything outside of their wholeness, but they produce these things by themselves, so too their eternal motion is not an affection but an activity. And here is a different argument. Since the heaven is a divine body and the activities of bodies are motions, it is necessary that the divine body have an eternal motion. Next, there is what the great Syrianus[175] proposed; he says that immortality is the activity of the cosmic god because, having reverted towards intellect and seeing that it is immortal and eternal, it yearns for immortality in itself, and, gaining its end and its own good, it actualises this immortality by moving eternally.

(Ad (i)) I think it is clear that one should also resolve the first of the matters of inquiry on the basis of this consideration. For if something has a certain activity as end and as good, it is for the sake of the activity and the function as its completion, just as immortality is the activity and completing function of the divine animal, and it is for the sake of immortality, in order that it be a divine animal. For each thing is in that in which its being is, and in the case of the divine animal its being is in eternal life and immortality. And the sciences which bring to completion are activities, and it is true to say that the scientist has as his end being a scientist in actuality. But arts have their activities in two ways. One is internal and corresponds to a positive condition and to being in actuality; and so artists <of this kind> could be said to be for the sake of knowledge, since each thing is because of its completion and, at least, its own good. But those arts which aim outside <themselves> also proceed on the basis of desire; and the person who is the kind of artist who looks outside <himself> could be reasonably said to be for the sake of the activity and also for the product.

286a12-31 Why, then, isn't the body of the heaven as a whole like this? [Because it is necessary that some of the body which moves in a circle, namely that at the centre, remain fixed.

But no part of it is able to remain fixed either at the centre or anywhere else. For <if it could remain fixed at the centre> its natural motion would be towards the centre; but it naturally moves in a circle, since <if that motion were not natural> it would not be eternal, since nothing unnatural is eternal. But the unnatural is posterior to the natural, and in the domain of coming to be the unnatural is a kind of departure from the natural. So, it is necessary that there be earth, since this rests at the centre. Let this be assumed for now; it will be proved later.[176]

286a22 However, if there is earth, there must also be fire, since if one of two opposites is by nature, it is necessary that the other be by nature, if, indeed, it is an opposite and it has some nature. For opposites have the same matter, and the affirmation is prior to the privation (I mean, for example, that hot is prior to cold), but rest and heaviness are said as privations of lightness and motion.

286a28 However, if there is fire and earth, there must also be the bodies between them, since each of the elements stands in opposition to each. And let this be assumed for now;] we must try to prove it later.[177]

Having proved that, being a divine body, the heaven moves with an eternal and consequently circular motion, he now inquires why the whole of the body of the heaven isn't like this, that is why it doesn't move in a circle with a single and simple motion. He is now calling the cosmos heaven.[178] And he answers, 'because it is necessary that some of the body which moves in a circle, namely that at the centre, remain fixed'. For what moves in a circle in the strict sense always contains something which does not move at the centre, about which it can move and which remains fixed. For it is also universally true that if something is going to move locally, it is necessary that some body from or around which it moves remain fixed, as has been proved in *On the Motion of Animals*.[179] For if someone were to say that it can move about its own central point, he would appear to say what is impossible, since the central point, being an incorporeal limit, cannot remain fixed when the things of which it is the limit move, since the central point is not something which exists *per se*. On the other hand, it is true that if the central point did not remain fixed, the universe would not revolve in the same <position>.[180]

Alexander also mentions this interpretation of the words 'the body of the heaven as a whole', which takes them to refer to the cosmos; however he prefers to understand the body of the heaven as a whole

as the body which moves in a circle, so that <Aristotle is saying that> something remains fixed in the centre of this, about which it moves. However, in the case of the motion of other animals, what remains fixed is some part of what itself moves,[181] and in fact Aristotle himself says here that it is necessary for some of the body which moves to remain fixed at the centre. However, the earth is a part of the cosmos and not, in fact, a part of the heaven. And <Aristotle> also recognises this; he says, 'but no part of it is able to remain fixed', meaning by 'it' the body which moves in a circle, not the cosmos about which the first things were said.[182] And <Alexander> argues in many ways that the earth is not a part of the heaven, even though he thinks that what remains fixed in the centre is a part of what moves in a circle. And Nikolaos the Peripatetic,[183] paraphrasing what is said here in his 399,1
work *On the Philosophy of Aristotle*, has set out the text this way:

> Why, then, is not the whole cosmos this way? Because it is necessary that something around the centre of what moves in a circle remain fixed. But the fifth body could not remain fixed nor be at the centre.

And Alexander himself correctly understands Aristotle when Aristotle proves that what remains fixed at the centre is not some part of the <body> which moves in a circle; for none of this can remain fixed anywhere else, since eternal motion belongs to all of it. And still more it cannot remain fixed at the centre; for if something naturally remained fixed at the centre, it would also move there naturally (for if a thing remains fixed in a place naturally, it also moves there naturally), and this would be its own natural motion. But there is one natural motion for each of the simple bodies; and for the heaven, since it is eternal, this is natural motion in a circle, so that motion towards the centre is not natural for it. Therefore resting in the centre is not natural for either the whole of it or part of it. Nor does it remain fixed unnaturally, since this body which moves in a circle has its motion eternally, but the unnatural is not eternal; for the unnatural is posterior to the natural, because in the domain of coming to be the unnatural is a departure from the natural, but what is a departure and a by-product of something is posterior to that from which it departs. And, in general, if what is unnatural is in <the realm of> coming to be and if there the natural is not eternal either, how could the unnatural be eternal? For even if the unnatural in coming to be were always, as the natural is, it would be different at different times and not always the same in the way that the heaven is naturally always the same in number.[184] Accordingly it is necessary that what is at the centre, since it is no part of the eternal body, be not eternal but destructible, heavy, stable, and consequently cold and dry. For destructible things are characterised by these passive qualities; earth is heavy and cold and dry. And <Aristotle> has <in effect> already proved that earth is in the centre, since he has proved that the centre is the appropriate place for what is heavy,[185] but earth is heavy; and he will prove it at the end of this book.[186]

(286a22) He says, 'however, if[187] there is earth, there must also be fire, since if one of two opposites is by nature, it is necessary that the other be by nature'. For the same matter is naturally related to both opposites in the same way. So if fire and earth are characterised by active opposite qualities, fire by heat, earth by coldness, they will be characterised even more by their local motions, since (i) fire moves towards the perimeter, earth towards the centre, (ii) natural motion is most of all substantiating for natural bodies, and (iii) natural bodies which have opposite <natural> motions have opposite natures. Consequently, even if water, being cold and moist, is opposite to fire with respect to both qualities, but earth is opposite to fire only with respect to cold, there is nothing surprising if earth has come to have a greater opposition to fire. For the greater denseness of earth makes it heavier and separates it further spatially <from fire>. For even if fire and earth are each said to be dry, nevertheless the form of dryness is completely different in each, since fire is light, moving,[188] and fine, <while> earth, <is> heavy, dense, and lifeless.

Having proved that if there is earth there must be fire on the basis of the antithesis of opposition, he proves the same thing on the basis of the antithesis of privation and positive condition. For if a privation exists, it is necessary that the corresponding positive condition exist in a prior way (he calls the positive condition an affirmation). And he now calls the naturally weaker of a pair of opposites the privation, so that cold is the privation of hot, and hot is the affirmation or positive condition corresponding to cold. So too rest and heaviness, by which earth is characterised, are spoken of as privations of motion and lightness, by which fire is characterised. So, if earth, which is a privation, exists, it is necessary that fire also exist and satisfy the definition of a positive condition; and it is clear that it is necessary that fire be prior by nature, since a positive condition is prior to its privation and an affirmation to its denial. He has clearly indicated that cold, heaviness, and rest are the things in terms of which he most understands the opposition and privative character of earth relative to fire. For fire is always moving, since when it is down it always moves up, and when it has come to be up, it is carried around with the divine body. And he says[189] that in general the more active elements also satisfy the definition of form and positive condition, the more passive that of privation and matter.

(286a28) But if the extremes, earth and fire, exist, it is necessary that the intermediates, water and air, also exist because each of them stands in opposition to one of earth and fire with respect to both of its qualities.[190] For earth, which is cold and dry, stands in opposition to air, which is hot and moist, and fire, which is hot and dry, stands in opposition to water, which is cold and moist. And the intermediates are also opposite to the extremes in their motions. And he has said that if one of two opposites exists, the other exists also because opposites have the same matter.

Alexander says – and I think correctly – that it is now not only possible to say that the intermediates are opposite to the extremes, but one can say this about all the elements. For whichever one you take, you will find that it stands in opposition to the other three. But if they all naturally stand in some opposition to the others, it is necessary that they all exist. And, in fact, if one were comparing two <of the elements> with two, I think it would not be appropriate for the accurate interpretation of Aristotle to say that one stands in opposition to one, but that each stands in opposition to each. 401,1

Having now proved the necessity that earth exists from the need for some unmoving centre for the body which moves in a circle and the necessity of the other elements from opposition, he nevertheless says that these things will be assumed for now but he will prove them later, since he will also have a lengthy discussion of the fact that it is necessary that these are the elements and that there are no others besides these in the fourth book[191] and in *On Coming to be and Destruction*.[192]

In the *Timaeus* Plato demonstrates that there are four elements in the cosmos[193] as follows:

> Since the cosmos has come to be, it is perceptible; therefore it is visible and tangible, visible because of fire, tangible because of earth; but since these two solids are opposites, if the cosmos is going to be solid and not plane, there is a need for two mean proportionals to fill out the solid proportionality.[194]

286a31-b9 Since these elements exist, [it is evident that there must be coming to be because none of them can be eternal; for they undergo and produce opposites in their interaction and they are destructive of one another. Furthermore, it is not reasonable that something movable for which it is not possible that the natural motion be eternal be itself eternal. But these things have a motion.[195] It is then clear from these considerations that there must be coming to be.

286b2 But if there is coming to be, it is also necessary that there be one or more other motions, since it is necessary that the elements of bodies relate to one another in the same way with respect to the motion of the universe. We will speak about this more clearly in what follows.[196]

286b6 For now this much is clear: for what reason there is a

> plurality of circular bodies, namely because there must be coming to be; and that there is coming to be since there is fire; and that there is fire and the other elements since there is earth; and that there is earth because it is necessary that there be something which always remains fixed] (since it is necessary that there be something which always moves).

Having proved that there must be four bodies standing in opposition
to one another, he says that it is evident that since these exist it is
necessary that there be coming to be and destruction. For if these four
elements are opposites, but opposites are not eternal since they come
20 to be from and are destroyed into one another, it is necessary for there
to be coming to be and destruction. And he proves that these things
come to be and are destroyed from their motions. For if the natural
motion of something is not eternal, it too will not be eternal. He says
that each of the natural bodies has its being in its appropriate motion,
so that the life of the divine body is also an eternal motion. But the
25 four elements, which are movable and move in a straight line, do not
have eternal motions. For he proved in the eighth book of the *Physics*[197] that no motion in a straight line can be eternal. And he makes
clear that eternal life is not an eternal motion of absolutely all
things,[198] but of things which have bodies and are movable when he
30 says here, 'but these things have a motion', since if they were not
movable it would not be necessary that they be completely destructible, even if they did not have an eternal motion. For what is motionless is also indestructible and has an eternal life, but those moving
things which have an eternal life also have a completely eternal
402,1 motion. When he says 'but these things have a motion', he leaves out
the words 'and the motion is not eternal'.

The preceding is the way Alexander interprets this passage. He
recognises – I think correctly – that eternal life is not also said by
5 Aristotle to be an eternal motion in the case of absolutely all things,
but only in the case of those that move. But he treats the word
'movable'[199] superficially; for he seems to understand it as equivalent
to 'moving' since he also seems to accept that Aristotle's words 'it is
not reasonable that something eternal … be itself movable' apply to
what moves eternally. But perhaps by 'movable' he means what
potentially has motion at some time; but what sometimes rests is of
10 this kind, and what moves in a straight line sometimes rests. And
Aristotle explains what he means by 'movable' when he adds 'for
which it is not possible that the natural motion be eternal'.[200] For if
the motion were eternal, the thing would not be movable, but moving.
And perhaps too the words 'but these things have a motion', have not
been written elliptically, but mean 'but these things move at some
times but not always', that is, these things are movable and do not

have an eternal natural motion, and therefore it is not reasonable that they be eternal; rather they come to be and are destroyed.[201]

(286b2) Having proved in this way that since the circular motion is eternal it is necessary that there be coming to be and destruction, he adds that it is necessary that there be one or more other motions. For these things change into one another in accordance with the motion of the divine bodies and a certain relation those bodies have to things which come to be. And if there were only one motion, the condition of things in our world would always be the same, since the affection which accrued to things here would always be one and simple. Consequently, leaving out the invisible activities of heavenly things affecting what is here, the activities of the sun and moon, which clearly change sublunary things in this way, would always be the same. For if the motion of the fixed sphere were one and the sun and the moon were fixed in the fixed sphere and moved about with it, there would be no difference among winter, summer, and the seasons in between. Nor would there be any difference in the daily change if the sun always travelled on the same course with the fixed sphere. For if the sun happened to be fixed in Cancer, our condition would always be summer, but if in Capricorn, it would always be winter, and there would be no coming to be and destruction, and no difference in the illuminations of the moon either. But if someone were to hypothesise that the sun moves on its own through the centre of the signs of the zodiac and the moon obliquely to the zodiac – since they are observed to proceed from Aries to Taurus and from Taurus to Gemini – he would first need a plurality of motions, as Aristotle says, and then also reverse motions. And he announces that he will speak more clearly about this in the sequel.

(286b6) 'For now', he says, 'this much is clear: for what reason there is a plurality of circular bodies', which is what we first proposed to investigate. Moving briefly from below and rising upward from consequents to antecedents, he brings together analytically the premisses of the demonstration. For there are many circular bodies which move in reverse because 'there must be coming to be'. And there is coming to be since there must be fire which stands in opposition to air and water (and that is why he also introduced them). And it is necessary that there also be fire and the other elements, since there is earth. And it is necessary that there be earth, because it is necessary that there be something which always remains fixed. And this is so since there must be something which always moves. He has put this last part together briefly.[202] The sequence <of argument> seems to be this: it is necessary that something remains fixed in the centre, if it is necessary that something always moves in a circle, and the latter is necessary if it is necessary that something always moves (since circular motion is the only eternal motion, since it is the only continuous one).

<Aristotle> says, 'it is necessary that there be something which always remains fixed since there must be something which always moves', and thereby makes the argument more universal. For, as he proved in his work on motion,[203] every motion, and not just circular motion, takes place about something which remains fixed. In the *Physics*[204] he inferred the necessity of there being something which is always in motion from the necessity of there always being motion (since it is not possible to assume a starting point for either time or motion). But if it is necessary that there always be motion, it is also necessary that there also always be something which is moved; for motion is in the thing moved, as he proved in book 3 of the *Physics*.[205] And this would be the method of demonstrating that there is an eternal body based on motion and taken in a way appropriate for discussions of motion.

But now he is discussing the heavenly and divine body, and he has based his demonstration on this. He says that it is necessary that the divine body move eternally, since it lives an eternal life; but this is necessary since it is necessary for it to be immortal; but this is necessary since it is necessary for what it is to be a divine body to belong to a divine body because of immortality in actuality.[206] Consequently, if you wish to synthesise the argument again, you will say:

(i) The heaven is a divine body;
(ii) a divine body is immortal in actuality;
(iii) what is immortal in actuality has an eternal life;
(iv) a body which has an eternal life has an eternal motion;
(v) it is necessary that what moves eternally move in a circle;
(vi) if it moves in a circle, it is necessary that there be a centre, some body which naturally remains fixed in the centre and the parts of which move naturally to the centre, being heavy and cold;
(vii) earth is of this kind;
(viii) but if there is earth, it is necessary that there be fire and the intermediates which are related to one another as opposites;
(ix) if these things exist it is necessary that there also be both coming to be and destruction;
(x) if there is coming to be and destruction, it is necessary that there be more than one heavenly motion and that they be in reverse of one another.

Obviously we infer this last necessity from <sensory> evidence; for
404,1 the motions of the heaven are causes of coming to be and destruction, and coming to be and destruction are <sensory> evidence for the motions; and some of the other conclusions in the argument were inferred from causes, some from <sensory> evidence.

Alexander says:

> In this passage Aristotle may also have been speaking about foresight concerning things in our world. For in this way both the change of the elements into one another and the coming to be of living things follows from the complex motion of the gods. But I raised the question whether, if the variegated motion of the divine body were eliminated by hypothesis but the four elements existed, they would change into one another or not. For if they didn't change, they would be eternal, which is impossible. And, in fact, if earth were eternal, there would be no necessity for the variegated motion of the gods, at least if the motion occurs because of the necessity that there always be earth. But if the elements did change into one another, since they are destructible by their own nature and the change of things which are destroyed is into opposites, and these are opposites, why would the variegated motion of the divine body be necessary for their coming to be and destruction?

Having raised this difficulty, he solves it in these words:

> Perhaps the motion of the gods is not the cause of the destruction of the elements and their changing into one another, but of these things happening in an ordered way. It is as a result of this that everything is also always preserved in terms of form, since insofar as it was a question of their own nature, nothing would prevent all of the elements from changing into one thing at some time, as those who believe in conflagration say,[207] and again changing from one thing into something different; for it is necessary that opposites not exist together in the same thing in the way that relatives can, nor has this been proved. However, it is necessary that they be by nature, but it is possible for things which are by nature to exist in turns, and if this happened, it would be possible that earth not exist at some time, if everything happened to change into fire. But if there were no earth, the circular motion would not exist. Therefore the variegated motion of the divine body is the cause of things always remaining the same in order and changing into one another in a measured way. For the[208] existence of living things is a consequence of the well-ordered change of things in our world and of their proper mixture.

Using these very words, Alexander says these things in answering his question. But, if it is necessary to tell the truth, I am, as the proverb puts it, a kind of 'white measure'[209] for investigations of this kind. For looking for the attributes of products while eliminating their causes in the strict sense seems to me to be more the function of play than of striving seriously. It is as if someone were to eliminate for the sake of argument the source of a river and keep the river and then to inquire whether or not the river which comes from the source would flow when the source had been eliminated, and then say that if it didn't flow, it would be absurd to call it a river, and add that if it did flow, the river would have no need of the source for its existence.

To say that the original hypothesis – that the river should remain when its source had been eliminated – is absurd is easy. But that the
405,1 four elements should remain when the motion of the heaven, which causes them to come to be, is eliminated is even more absurd. As a result Alexander's investigation is off the mark in my opinion and his resolution of it – that the heavenly revolutions are not the causes of the change of the elements – seems to me out of tune. For who doesn't know that when the sun is near to our zenith, it heats things around us and turns liquid things into vapour and makes air and fire come to be from water, and when it is at a distance, it aggregates the things which have been rarefied again and makes water come to be? But let the <heavenly motions> be the cause of the fact that change of these things is ordered and measured in a determinate way. It is still necessary to say that, prior to this, one ought to make nothing other than the heavenly motion responsible for change itself in general and for the existence of the things which come to be as a result of the change.

And Alexander himself notes this, since he adds a second answer[210] to the question, which is the following:

> If everything that comes to be not only comes to be from something but by the agency of something (since things don't just come to be from matter but by the agency of an efficient cause), the elements by themselves will not be sufficient <to bring about> their change into one another and coming to be from one another; for they need something to change them, since they relate to one another as matter.

However, a person who knew such things should have rejected the hypothesis of the inquiry straightaway because it hypothesised that the four elements exist and either change into one another or are eternal, if the heavenly motion is eliminated. And how can <Alexander> say that it is necessary that opposites not exist together in the same thing, but that it is possible that they exist in turns, since it is possible that the earth not exist at some time if everything happened to change into fire and again from fire into something else? For, first of all, if it were possible, the possibility would <in fact> occur in infinite time, since the elements always exist; and then, if what changes does not only change from an opposite but by the agency of an opposite into which the change takes place, then, if there were nothing else but fire, what would change it so that something else came to be from it? But, since I understand these views of Alexander and more, I should take up what comes next.

[Chapter 4]

286b10-287a11 It is necessary that the heaven have a spherical shape, [since this shape is both most appropriate for its substance and it is naturally first.[211]

286b11 Let us discuss shapes universally and say which sort is first, both among plane figures and among solids. In fact every plane figure is either rectilinear or curvilinear, and what is rectilinear is contained by several lines, what is curvilinear by one. Since in every genus what is one is prior to what is multiple and what is simple to what is composite, the circle will be the first plane figure.

286b18 Furthermore, since something is complete if it is not possible to take anything outside it which belongs to it, as was determined previously, and it is always possible to add to a straight line, but never to the line of a circle, it is evident that the line containing a circle will be complete. Consequently, if what is complete is prior to what is incomplete, it will result that the circle is the first figure.

286b23 Likewise, too, the sphere is the first solid, since it alone is contained by one surface, whereas rectilinear solids are contained by several. For as the circle stands among plane figures so the sphere stands among solids.

286b27 Furthermore, even those who divide bodies into plane figures and generate bodies from plane figures seem to be witnesses for this, since the sphere is the only solid which they do not divide because it does not have more than one surface. For the division into plane figures is not carried out in the manner of someone cutting a whole into parts, but in the manner of someone cutting a whole into things different in form. It is clear, then, that the sphere is the first solid figure.

286b33 It is also most reasonable for those who assign <figures> a numerical ordering to do so by assigning one to the circle and two to the triangle (since it <has angles equal to> two right angles). But if the triangles were assigned one, the circle would no longer be a figure. But since the first shape belongs to the first body and the body in the last revolution is the first body, the body which revolves in a circle will be spherical.

287a5 So too, then, is the body continuous with it, since what is continuous with something spherical is spherical. And similarly for the things at the centre of these <spheres>, since it is necessary that things which are contained by spherical <bodies> and touch them as wholes be spherical. But the things beneath the sphere of the planets touch the sphere above them. Consequently all of it will be spherical, since everything touches] and is continuous with spheres.

After the discussion of the substance of the heavenly body, which he finished up in the first book, and having proved in the preceding that motion in a circle is natural and not forced and then having inquired into the spatial relations, first finding the relation of right and left on the basis of the starting point of motion and inferring the others from this, and then finding the cause of the contrary motion, which is clearer after the inquiries concerning motion, he next inquires about the shape of the heaven and demonstrates clearly and precisely and in an ordered way by means of several arguments that it is spherical.

The first argument proceeds from an axiom which says that the naturally first shape belongs to the naturally first body, and <from another axiom which says> that the heavenly body is the naturally first body. Taking these things as clear, he constructs the argument as follows: spherical shape is the first shape; the first shape belongs to the first body, i.e. the heavenly body.

(286b11) He proves that the first figure is the spherical figure universally, distinguishing among figures 'which sort is first, both among plane figures and solids'. Then, dividing plane figures into rectilinear figures and curvilinear ones, he proves that the curvilinear figure *par excellence*,[212] the circle, is both simpler and more complete than rectilinear figures. And he proves that it is simpler from definitions. For if (i) a circle is a plane figure contained by one line, (ii) the simplest rectilinear figure, the triangle, is contained by three straight lines, (iii) in every genus what is one is prior to what is many, and (iv) the simple is always prior to what is composite (since composite things are composed from simple ones), the circle will be the first plane figure.

(286b18) However, among plane figures the circle is also complete, since something is complete if it is not possible to take any of its parts outside it. And, as was specified earlier,[213] something is not complete if it is possible to add to it in an appropriate way; and it is always possible to add on to a straight line, but it is not possible to add to the line which contains a circle. So, rectilinear figures are contained by straight lines which are never complete, the circle by its circumference, which is always complete. It is clear then[214] that among plane figures the circle is complete. But what is complete is everywhere prior in nature to what is incomplete, since the incomplete acquires its share of completeness from complete things. For this reason, then, the circle is prior to rectilinear figures.

And the complete is prior to the incomplete not only in substance but also in time.[215] For when both are taken as amounting to the same thing as in the case of things which come to be, the incomplete is prior

to the complete, since the seed is prior in time to the fetus, the fetus to the baby. But the producer of the seed precedes the seed in time, and in the case of each thing that comes to be its generator[216] and its 407,1
eternal cause precede it – complete things <preceding> an incomplete one.

(286b23) And, just as among plane figures the circle is simple and complete, so among solids the sphere is simple and complete and for the same reasons. For among solids the sphere is contained by one surface, whereas rectilinear solid figures are contained by several, and only the surface which contains a sphere is complete, since it does not admit the appropriate kind of addition. For as the circle is among plane figures so is the sphere among solids. Therefore the sphere is the first solid figure in substance, and it is reasonably associated with the body which is first in substance.

(286b27) Those who generate bodies from plane figures and divide them into plane figures make clear that the sphere is the only simple solid figure; and the natural philosophers[217] had this conception of it. For, although these people say that plane figures are the elements and principles of bodies, nevertheless the sphere is the only solid which they do not divide 'because it does not have more than one surface'. How so? Someone might say, 'isn't a sphere a body, doesn't it have parts, and isn't it possible for it to be divided into its parts?' Aristotle dissolves this objection by saying that a division into plane figures is different from a division into parts. For division into plane figures proceeds into simpler and more elemental things which are different in form, since it is into incorporeals rather than into bodies. But a body which is divided into its parts is divided into bodies which have the same substantiality as the whole, as when a sphere is divided into hemispheres; and this kind of division into parts also attaches to simple things, but division into plane figures attaches only to composites. So the fact that these people do not divide the sphere into plane figures is evidence that they consider it to be simple.

(286b33) And he again proves that the circle is the first plane figure and the sphere the first solid in another way, namely from the order of numbers. For if one is prior to two and one is assigned to the circle because it is contained by one line[218] and two is assigned to the triangle (which is the simplest rectilinear figure) because it has interior angles equal to two right angles, the circle will be first. But if someone were to determine the number one as triangle, the circle, being simpler than the triangle, would no longer be a figure, since nothing is simpler than one. So, if as among plane figures the circle is related to rectilinear figures, so among solids the sphere is related to them, it is clear that the sphere will be first.[219]

Omitting this proportion on the grounds that it is clear, he adds the conclusion of the whole demonstration after the demonstration with a causal conjunction <'since'>. For since the first, i.e. (as has been demonstrated) spherical, shape belongs to the first body (since it is necessary that what is first belong to what is first), but the first body is the body of the fixed sphere – which he calls 'the body in the last revolution' – it follows that spherical shape belongs to the fixed heavenly body, the body which revolves in a circle. If he proved[220] earlier that the heavenly body is naturally first on the basis of its circular motion – for he proved on this basis that it is eternal – and he now proves that it is spherical on the basis of the fact that it is naturally first, no one should think that his proof is circular. For moving in a circle and being spherical are not the same thing, since it is possible for something to be a sphere and not move in a circle or to move in a circle and not be a sphere.

(287a5) Having proved that the body of the fixed sphere is spherical on the grounds that the first shape, the spherical, suits the first body, he proves as a consequence that the bodies which are contained by it and touch its concave surface at every point[221] are also spherical. For either it is necessary that there be a void in between, which has been proved to be impossible,[222] or, if everything is filled up, it is necessary that what touches the concave surface of a spherical figure at every point be itself spherical. And, in accordance with the same inference, the things after the divine body which come to be and are destroyed (all of which he now designates as one thing 'at the centre') are also spherical. For it is necessary that things contained by what is spherical and touching it at every point so as to leave behind no void also be spherical, but what is under the moon touches the lunar sphere.

He has now proved that the shape of the whole body which comes to be, taken as a single thing, is spherical. And, proceeding a little further, he will demonstrate that the appropriate shape for each of the four <elements> is spherical.[223] For even if the[224] external surface of fire is spherical because it touches a spherical surface <at every point>, it is not necessary that what is inside fire be similar. It is necessary in the case of things which have a spherical shape because of their own substance but not in the case of things which are given shape by other things.

This is as far as he has filled out the first argument proving that the heaven is spherical on the basis of the necessity that the first shape belong to the first body because it is suitable to its substance and nature; and if the heaven is spherical, so is the whole cosmos. It would be good also to hear what Plato says about the shape of the cosmos:

> <The demiurge> gave it a shape which was fitting and akin to it. And the fitting shape for the living thing which was going to contain all[225] in it would be one which contains in itself all

> figures whatsoever.[226] Therefore he, considering uniformity to be very much better than non-uniformity, turned it out[227] round and spherical with an equal distance from its centre to its[228] extremities in every direction, <making it> the most complete 409,1
> figure and the one most similar to itself.[229]

What then is the difference between saying 'fitting and akin to it' and saying 'most appropriate for its substance and its nature'?[230] And <Plato> has also said that the spherical shape is the most complete of all shapes.

And in what comes next, Aristotle will apparently fail to mention other things.[231]

> **287a11-22** Furthermore, since the universe appears, and is assumed, [to revolve in a circle and it has been proved that there is neither void nor place outside the last revolution, it is necessary also as a result of these facts that it be spherical. For if it were rectilinear, it would result that there was also place and body and void outside. For something rectilinear turning in a circle will never occupy the same space; rather where there was a body previously, there will not be one now, and where there is not one now there will be one again, because of the change of position of the angles.
>
> 287a19 Similarly, even if it were to be some other shape not having its lines from the centre equal, e.g. if it were shaped like a lentil or egg. For in all <such cases> it would result that there is place and void outside its revolution,][232] because the whole does not remain in the same space.

This is the second argument; it proves that the heaven is spherical on the basis of what has been demonstrated already. If the heaven revolves in a circle (as has been proved before)[233] and there is neither void nor place outside it (and this also has been demonstrated already),[234] it is necessary that it be spherical. For, if its shape were rectilinear, in revolving it would never remain in the same place, since its parts would not move into one another's locations; for the extrusions of the angles when they moved would make it turn out that where there was a body previously there is not one now, and where there is not one now there will be one again, since they will take up a greater place than the other parts of the rectilinear figure, with the result that outside there would be void and place which received the extrusions of the angles.

The words 'it would result that there was also place and body and void outside' may mean 'it would result that there was either place or body or void outside' (since there would always be one of these things outside the heaven)[235] or rather that there would be all the things which have been demonstrated not to exist outside the heaven – that is, body, void, place – if the heaven moved in a circle and had a rectilinear shape. For if 'where there was a body previously there will not be one now', there would be void, and if 'where there is not one now there will be one again', there would be body and also a place to receive it, since what was previously void would become a place containing body. A body which came to be outside it would be nothing other than parts of it, but insofar as the part outside it had not been a body, a body would come to be <outside the cosmos> in the revolution.

(287a19) He says that if the shape of the heaven isn't rectilinear, but is something other than spherical (this is what he indicates with the words 'not having its lines from the centre equal, e.g., if it were shaped like a lentil or egg'), 'in all such cases it would result that there is place and void outside its revolution because the whole does not remain in the same space'.

These things which Aristotle says are also well criticised by Alexander. He says that the argument is not completely sound in these cases; rather, if the heaven were shaped like a lentil and had its poles lying in the central points of the planes,[236] in revolving it would always remain in the same place, but if it had its poles lying in the
410,1 circumference so that its revolution described a circular figure, it would be necessary that it leave behind void and place. Similarly, too, if it were shaped like an egg and had its poles not through the length <of the egg> but sideways so that the rotation occurred relative to the length it would be necessary that it leave void behind; but it would not be necessary if it had its poles through the length. And also in the case of a cylinder and a cone and the solid called rhomboidal, which is composed of two cones connected at their bases, if they move around their axis they always remain in the same place.[237] Why then, when things are indeterminate in this way, has Aristotle asserted this? Perhaps understanding the words 'it would result' in a strict sense solves the difficulty. For in all these cases, although the poles can be taken in different ways, if the poles are taken in a certain way, it would result that there is place and void outside its motion; but in the case of a spherical figure because of its uniformity, however the poles are taken, it will always rotate in the same place because it has all <its straight lines> from the centre equal. For, since it is homoiomerous, why would it stand at a greater or less distance from the centre in this direction or that?

So Aristotle produces the following syllogism in the second figure to the effect that the heaven is spherical based on what has been demonstrated already using physical considerations, namely that there is no body, place, or void outside the heaven:

(i) The heaven, moving in a circle, always remains in the same place;
(ii) none of the things having a shape other than spherical which move in a circle always remain in the same place, but revolving rectilinear figures leave behind place and void and body, as was said previously, and all curvilinear figures do not always fill their place (since it results from some way of taking their poles that they leave behind and come to occupy external place);
(iii) therefore, the heaven can have nothing other than a spherical shape.

The astronomers, who agree that some figures – such as the cylinder, the cone, the lentil-shaped, the egg-shaped, and the solid called rhomboidal – always remain in the same place when their poles are taken in a certain way, prove on the basis of the phenomena that the heaven can have no other than a spherical shape. For <if the heaven had one of these other shapes>, then, although the fixed stars, which are thought always to move on the same parallel circles, would maintain the same distance from us, the stars which are said to wander and which move on an oblique circle would pass from one parallel circle to another and would have a greater and a less distance from us, not because they reach apogees and perigees, as we now say, but because the straight lines from the earth falling on the different parts of the heaven would be unequal in the case of the figures which are other than spherical, so that these <planets> will also appear to be at greater and less distances from each other and from themselves according to the different distances which they have from us, even though they are on the same surface.[238] Ptolemy also inferred that the shape is spherical from 'the impossibility that constructions of sundials harmonise on any other hypothesis'.[239] Expressing himself in a physical way, that wonderful man <Ptolemy> has recalled things said by Plato.[240] He says, '*aithêr* is the finest and most homoiomerous of all bodies, and the surfaces of homoiomerous bodies are homoiomerous, and among plane figures only a round figure <has homoiomerous surfaces> and among solids only a spherical figure has them'.[241]

287a23-30 Furthermore if the motion of the heaven is the measure [of motions because only it is continuous, uniform, and eternal, and what is least in each thing is its measure, and the fastest motion is the least, it is clear that the motion of the heaven will be the fastest of all motions. However, the line of the circle is the least line which goes from <a point> to itself.[242] But the fastest motion will be along the least line. Consequently, if the heaven moves in a circle and moves fastest,] it is necessary that it be spherical.

This is the third argument proving that the shape of the earth is spherical. The construction of the argument is this:

(i) The motion of the heaven is the measure of motions;
(ii) a measuring motion is a least motion;
(iii) the fastest motion is least;
(iv) of motions proceeding from the same power[243] that which covers the least distance is the fastest;
(v) of lines containing equal plane areas the least is a circular line,[244] and for solids a spherical surface is the least;
(vi) therefore, the heavenly revolution belongs to a body which has a spherical shape.

Of the premisses taken individually the one <stating> that the heavenly revolution is the measure of motions has been proved in the fourth book of the *Physics* in the discussion of time,[245] and he now recalls the reason briefly <by saying> because only it is a continuous, uniform, and eternal motion. For it is necessary that the measure of motions be itself also a motion, as the measure of magnitudes is a magnitude, that of monads is a monad, that of pints is a pint,[246] and that of weights is a weight. And it is also necessary that the motion measuring others be continuous; for if it were divided by a rest period, how could it measure continuous things? And it is also necessary that the measure be uniform. For if it increased or decreased, how could it be a measure? It is as if someone were to use his forearm as a measure, sometimes extending it and sometimes contracting it. And it is necessary that the measuring motion be eternal; for, since there is always motion, it is necessary that what measures it also always exist. Together all of these things, continuity, uniformity, eternity, belong only to the motion of the heaven. It is, therefore, correct to say that the revolution of the heaven is the measure of other motions.

In the case of every genus of things measured or measuring, the least is the measure. For if it is stades, a stade is the measure, if monads, a monad, and if pints, a pint.[247] And if something else is a measure, it still is also measured by the least; for the forearm is a measure, but it is nevertheless measured by the foot, and the foot is measured by the finger.

The least motion is the motion which occurs in the least time, and what makes a revolution around an equal <area> in the least time is the fastest motion. But a motion which makes a revolution around an equal area (e.g. an area of one hundred thousand square feet) through the least distance containing it is the fastest motion proceeding from the same power, so that, if the motion which goes around the same area remains the same, the motion will at the same time be fastest in terms of time and least in terms of the distance containing <the area>.

The least <distance> which goes from <a point> to itself, that is, the least of the distances containing and delimiting some figure, is the circular in the case of plane figures and the spherical in the case of solids. For this was proved certainly even before Aristotle[248] – since he has made use of it as something which has been proved – and <it was proved> more generally by Archimedes and Zenodorus[249] that of isoperimetric figures the most spacious is the circle among plane figures and the sphere among solids. What Aristotle says now, namely, that of lines containing equal plane figures the circular line is least (and similarly for the sphere in the case of solids) is a consequence of this; for <in cases when> if there are equal containers, the contents <of one> is less, if the contents are equal the container <of that one> will be less.[250] Therefore, it is necessary that the fastest and least motion is circular in the case of plane figures and spherical in the case of solids.[251] For if there were some other figure having <a volume> equal to that of a spherical figure, then, since in the case of equal bulk the surface of a sphere is less than that of differently shaped things, the motion of the sphere will be less when the power is the same. Therefore, the motion of the differently shaped thing will be greater, which will be absurd, if it is going to be the measure of motion. For each thing which moves in a circle moves along its perimeter, just as each thing which moves in a straight line moves along the underlying interval.

In connection with this text Alexander asks how Aristotle can say that of equal plane figures the circle is contained by the least line if the quadrature of the circle had not yet been discovered in his time. For, he says, the comparison of the line of the circle with straight lines depends upon the quadrature of the circle.[252] He answers this question by saying:

> Perhaps when these things were written the quadrature had 413,1
> been discovered to a certain extent. For even now it is not agreed that it has been discovered because the proofs put forward have proceeded on the basis of hypotheses. Or perhaps he is not now comparing the circle with rectilinear figures but with curvilinear ones such as the so-called ellipse. For the circle has the least perimeter in comparison not only with equal rectilinear figures but also with curvilinear ones. And similarly the sphere has a lesser surface not only than rectilinear solids but also with those which are contained by one surface, such as the lentil-shaped and egg-shaped, when they are equal <in volume>.

Alexander confirms that Aristotle makes his comparison in terms of these things on the basis of the fact that the words 'from <a point> to itself'[253] are more suitably said of figures contained by one line or surface.

However, in proceeding a little Alexander himself asserts that this is said in order to indicate the perimeter of a figure, since ‘a figure is what is contained by one or more boundaries’.[254] He says that the lines which contain things of this kind go from <a point> to itself;[255] for, he says, a straight line is not a perimeter, since it does not go from <a point> to itself,[256] and neither does an angle, so that it isn’t a figure either. Consequently <Aristotle> has said this about figure in general and not just about curvilinear ones.

Alexander also tries to prove in another way that Aristotle is now making a comparison with curvilinear figures. He says that the preceding argument[257] which invoked a void was evident in the case of rectilinear figures because of the extrusions of the angles, but it is not the same in the case of curvilinear figures. Consequently he might have added this argument to cover things which were not shown in the same way by the previous proof.

But perhaps neither Alexander’s question nor his answer has any use for what we’re dealing with, nor was he correct to say that the comparison of the line of a circle[258] with straight lines depends upon the quadrature of the circle. For the person proposing to square a circle is not proposing to find four equal straight lines equal to the single line of a circle from which the square might be produced; for this was found long, long ago and from it the line of a circle was found to be three and a seventh times its diameter, and there was no difficulty in producing a straight line three and a seventh times the diameter and hence equal to the line of the circle, and dividing it into four equal straight lines from which the square <with perimeter equal to that of the circle> is constructible. But the problem which is
414,1 formulated as ‘to construct a square equal to a circle’ demands a square plane figure equal to a circular plane figure, just as when a teacher of geometry[259] proposes to construct a square equal to a given triangle he is not proposing to make the lines of a square equal to the lines of a triangle, but an area equal to an area.[260]

So when Aristotle says that ‘the line of the circle is the least line which goes from <a point> to itself’,[261] he is comparing lines with one another, and the person who says that the area of a circle is greater than the areas of other isoperimetric figures is hypothesising that the equal lines containing the figures have been found and saying that the area of the circle is greater than both the areas of the other figures contained by a line equal to the perimeter of the circle and <the area of> the square <contained by four lines together equal to the perimeter>. He is not saying that the equivalent of the construction of the areas has already been apprehended.

One should recognise that this is also a Platonic argument which Aristotle has embraced. For if the line which contains a circular area is the least of those containing equal areas and least because the circle is more spacious than other isoperimetric figures – as Plato makes clear by saying that it contains all figures whatsoever[262] – then, even if Plato indicates something else with these words, this <argument> has also been taken from him.

> **287a30-b4** One could also adopt this belief [on the basis of the bodies situated around the centre <of the cosmos>. For if water surrounds earth, air water, and fire air, the upper bodies will also be in the same relation; for they are not continuous with these others, but they touch them. But the surface of the water is spherical, and what is continuous with the spherical or surrounds the spherical must also be spherical. So, it will also be evident from this] that the heaven is spherical.[263]

This is the fourth argument; it proves that the heaven is spherical from the fact that water is spherical, and not only that the heaven is spherical but that the elements above water which come to be and are destroyed, namely air and fire, are so as well. He will demonstrate later[264] that the earth is spherical with respect to perception.[265] And the argument here is the following:

> If the surface of water is spherical, and it is necessary that what is continuous with something spherical or touches it everywhere be spherical (and the latter is the way air is related to water because there is no void between them), air too will be spherical. For the same reason fire, which contains air, will be of this kind, and the things which come next and touch what is beneath them will be in the same relation right up to fixed sphere.

Alexander objects to this demonstration. He says:

> He might prove that the inner surfaces are of this kind with this argument, but not the outer ones. But since the aitherial body is also spherical, the other bodies down to water will be spherical, if they have their outer surface spherical because of what lies above them and their inner surface spherical because of what lies below. And perhaps, as far as these arguments <in the present passage> go, neither the outer surface of air nor the
> inner surface of fire has any necessity of being spherical. Or is 415,1
> it necessary? For air, which is liquid, flows like water into concavities and makes its outer surface spherical, as water does too. But if this is the way things are, the inner surface of fire will also be spherical, but its outer surface <will also be spherical> if the divine body is spherical, as was proved previously <in this chapter>. So, through the two proofs it would be shown that the intermediate bodies have each of their surfaces spherical, the

outer surface because of what contains them, the inner surface because of what they contain.

This is what Alexander says. But Aristotle proceeds as if the sphericity of the heaven was proved by this argument. He starts by saying that one could adopt this belief that the heaven is necessarily spherical 'on the basis of the bodies situated around the centre', and he finishes up the argument by saying, 'so, it will also be evident from this that the heaven is spherical'. How then can we say that only half of the demonstration has been obtained from this argument? Perhaps then this argument is in itself sufficient if we pay attention to Aristotle's precision. For if, because water is spherical, air has its concave surface, which touches the water, spherical, and it is made spherical outside because it flows everywhere around the water,[266] which is spherical, it is clear that the inner surface of fire will be spherical too. But its outer surface moves along with the fixed sphere from east to west, as the portents in fire make clear – I mean comets, meteors, and such things, which rise and set together with the fixed stars. So, since the body which carries the moon moves from west to east and fire touches that body everywhere, unless fire were smooth, it would prevent motion and be prevented from moving. As a result it is clear that the inner surface of <the body which carries> the moon[267] is spherical. But so is the outer surface. For again the moon's particular motion is alongside of the motion of Mercury, and this pattern proceeds up to the fixed sphere. In general, given that things under the moon have been proved to be spherical, it would be extremely anomalous if heavenly things had a worse shape than sphericity. So Aristotle is correct to say 'the upper bodies will also be in the same relation'.

287b4-14 But that the surface of water is spherical [is evident for those who make the hypothesis that water is of such a nature as always to flow into what is more hollow.[268] And what is more hollow is what is closer to the central point. So let straight lines AB and AC be drawn from the central point, and let the straight lines on which BC lies be joined. Then the straight line drawn to the base, the line on which AD lies, is less than the radii, so that the place is more hollow. So, let water flow around until it is made equal. But the radius AE is equal. So it is necessary that the water come towards the radii, since then it will rest. But the radius touches the periphery.] Therefore the surface of the water, on which BEC[269] lies, is spherical.

The preceding argument was made on the basis of the hypothesis which says that water is spherical. Now he proposes to prove this fact itself, and he introduces into the demonstration the precision of lines

and argues in a way which is really geometric in nature. He also 416,1
adopts a hypothesis here, but it is made very clear by the phenomena that 'water is of such a nature as always to flow into what is more hollow'. And always a place is more hollow if it is nearer to the space towards which water is so constituted as to incline and to which it flows if it is not prevented. He now calls this the central point, taking as a point what one might call the most hollow thing to which water flows. A little later it will be proved[270] that it is the central point of the surface of the water.

From this central point A he extends equal straight lines, AB and AC, and, joining BC, he produces a triangle ABC. And he draws AD from the vertex to the base between the points B, C. AD, then,

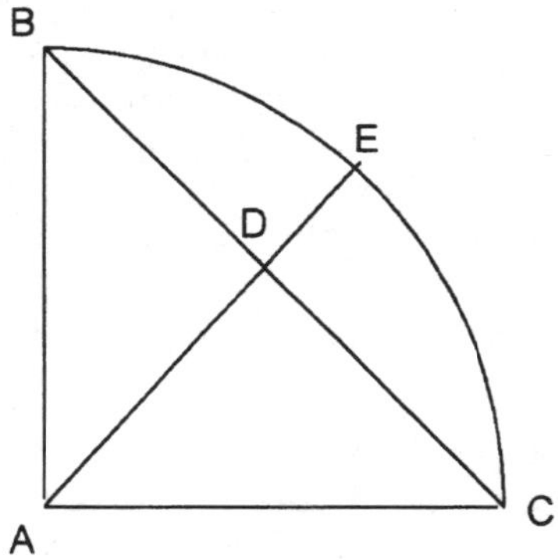

Figure 2 (416,8-417,6)[271]

is less than AB, AC; for if we draw a circle with central point A and radius either of <A>B, <A>C, then AB, AC will be radii which touch the circumference. But the straight line AD drawn between them will not touch it, and it too is drawn from the central point; therefore it will be less than they, so that D will be nearer A than B, C are. Therefore the place of D is more hollow relative to B, C, so that water will flow to D until AD has been made equal to AB, AC, and then D will no longer be more hollow than B, C. So if we extend AD and cut off AE equal to AB, AC, there will no longer be a more hollow place into which water flows. For if someone thinks that there are such places between the three straight lines we will produce straight lines from A through those points equal to the three <AB, AE, AC> and fill up the hollow. So if, in the case of the points A, E, C any other limits of equal straight lines drawn from the point A are found between them, then, if the straight lines from A are drawn in a plane, the line which is joined at the limits of the straight lines will be circular, and if they are drawn in a solid, what is extended through the limits will be a spherical surface. For if more than two equal straight lines are extended in the same direction from some point, if they are in the same plane, the line drawn through their limits will be circular, and if they are in different planes, the surface <passing through them> will be spherical. If, then, water naturally rests and does not flow only when it reaches the limits of the equal straight lines extended from A and what is drawn through those limits in three dimensions makes a spherical surface, then water has a spherical surface and its central point is the point from which the surface is everywhere equidistant. Therefore Aristotle has also called it a central point, whether this is the central point of the earth or of the universe or of water itself or whether all the central points occupy the same place although they have different significances.

Drops which fall into a soft place, such as the leaf of an herb or of a tree or of the pods of the Nile,[272] and rest after curling up and becoming spherical make clear that the surface of water is spherical. And still clearer <evidence> is the fact that sailors on the sea often do not see the earth or things that rise up in it from their ships, but when they climb the mast they see them as if they were near. But also if someone throws water and then coins or pebbles or some other lumps into a cup, the water will form a sphere and will not overflow until it surpasses the spherical surface.

> **287b14-21** It is clear from these things that the cosmos is spherical [and that it is turned out so precisely that nothing made by hands nor anything else we see with our eyes comes close to it. For none of the things from which it is constructed is capable of receiving uniformity and precision in the way that the nature of the surrounding body is. And it is clear that there is a proportion: as water is to earth,] so is it always with the more distant coordinates.[273]

The issue was to prove that the heavenly body is spherical. But since the whole cosmos (as far as the heaven is concerned) is spherical and the argument has proved that most of the things under the moon are spherical, it is not unreasonable for him to finish up his discussion by saying that the whole cosmos has been demonstrated to be spherical. But since he is in discussion with people who have become accustomed to artificial and natural spherical shapes, he also verifies in a completely god-loving way the superiority in shape of the heavenly body by comparison with all the others, and he verifies that it is not called spherical in an approximate sense (in the way we call the earth spherical with respect to perception,[274] even though it has such great mountainous protrusions and hollow ravines); rather it is a sphere which is turned out so precisely that nothing artificial or natural – or in general nothing[275] we see with our eyes – comes close to it.

But this extendedness of the discussion would require some exhortation. For even if it has been demonstrated to be spherical, the fact that its precision is not immediately apparent to perception produces disbelief. Therefore he proves that no other of the bodies in the cosmos from which the cosmos is constructed is capable of receiving uniformity and precision of shape in the way that the nature of the surrounding body is. He proves this on the basis of a proportion involving the elements which we see. For it is manifest to everyone that the surface of water is smoother and more uniform than the surface of earth. If then the elements stand in a proportion to one another so that as water is to earth, so is air to water and fire to air, and of coordinates – that is, of elements lying next <one above the other> – the one more distant from the earth exceeds the one under it in uniformity and in fineness and purity, it is clear that what is above fire also has fineness, purity, uniformity, and consequently precision of shape in proportion to its distance <from the earth>.

One should notice that the expression 'turned out' has been taken from the Platonic text which says:

> Therefore <the demiurge> turned <the body of the cosmos> out[276] round and spherical with an equal distance from its centre to its extremities in every direction.[277]

He has said that the cosmos is spherical because the heaven is spherical, and now he has made clear by what he has said that none of the bodies in the cosmos is capable of admitting uniformity and precision in the way that the nature of the surrounding body is.

[Chapter 5]

> **287b22-288a12** There are two ways of moving in a circle, [for example, from A either to B or to C, and it has been said previously[278] that these motions are not opposites. But, if it is not possible that anything eternal be a matter of chance or spontaneity, and the heaven and its circular motion is eternal, why does it move in one direction and not in the other? For it is necessary that this <direction of motion> either be a principle or that there be a principle for it.
>
> Perhaps trying to say something on some subjects or on all and to pass over none would seem to be a sign of great simple-mindedness or great zeal. However, it is not right to censure all people in the same way; rather one should look at what their reason is for saying <what they do> and further how the person stands with respect to trust,[279] whether it is a human affair or something stronger. So when someone hits on more precise necessities, one should show gratitude to the discoverer. But now we should discuss what appears to be true.
>
> 288a2 If nature always makes the best possible thing, and if,

> just as in the case of motions in a straight line that towards the upper region is more honourable (since the upper region is more divine than the lower one) and in the same way too motion forward <is more honourable> than motion backward, then, since <the heaven> has right and left (as was said previously), it also has – as the difficulty being raised also gives witness – priority and posteriority. For this explanation solves the difficulty. For if it is in the best possible condition for it, this will also be the explanation of what we have mentioned. For it is best to move with a simple and unceasing motion,] and this in the direction of what is more honourable.

He now proposes the fifth inquiry[280] concerning the motion of the first heaven: since the heaven moves in a circle it would be possible for it to move in two ways, either in the way in which it now moves or in reverse so that the fixed sphere, moving from what is now called the place of setting <i.e. the west>, would make its rising there and the planets obviously would move from what is now called the place of rising <i.e. the east>. And first he proves that it is reasonable to engage in this inquiry on the grounds that there is also some reason for this <direction of motion>. For if the heaven is eternal and moves eternally, and nothing eternal occurs pointlessly or spontaneously, it is clear that there is some reason why the motion is the way it is now and not in reverse. Since, then, moving in such a way has an explanation, it is either a principle and explanation of other things, as the contrary motion was said[281] to be the principle of coming to be, or there is some other principle of it, as the necessity of the first figure attaching to the first body is the principle <which explains> the sphericity <of the heaven>.

Next, considering the difficulty of the inquiry, he determines which people should look into issues of this kind: they should not be people who love glory or who rashly trust their own imaginations, but rather people who love learning and restrain their assent more prudently. It is suitable for these people to say something about some subjects which are thought to be most difficult and to try to inquire into all subjects and to pass over none. For in the case of these people the reason for making such inquiries will not be simplemindedness or excessive zeal.

He says that there are two ways of moving with a circular motion, one from A to B, the other from A to C, and he recalls that this does
419,1 not undermine the argument that there is no motion opposite to motion in a circle, since these two motions are not opposites, but motions in reverse directions. Recognising that the development of the argument about to be given will be suspect to many people and particularly to those who investigate on the basis of what is accessible, with philosophical prudence he makes a first defence of it by

saying that when someone encounters people who discover the more precise necessities of an explanation which is being sought, one should show gratitude to them, but, until we chance upon them we should in the meantime say what now appears to be true.

(288a2) <Aristotle> affirms that nature (and still more the demiurge) always makes the best possible thing come to be; and he affirms that, just as in the case of motion in length 'that towards the upper region is more honourable' than that towards the lower, in the same way too motion forward is more honourable than motion backward; and since when he proved previously that the universe has an up and a down and a right and a left,[282] he did not say anything about front and back,[283] but now he makes use of it as real, he reasonably demonstrates that <the universe> has these things since it also has right and left. For if motion is from the right and towards the front, and whenever that from which exists that towards which also exists, it is clear that if right exists so does front. For what starts to move moves towards something and if the starting point exists so does what is from the starting point.

He says that the difficulty being raised (that is, the difficulty being inquired into, why the heaven moves in a certain direction and not the reverse) gives witness that the heaven has a better and worse <aspect> in respect to the dimensions, and that front is better than back. And he adds how the difficulty gives witness for this when he says, 'for this explanation solves the difficulty'. For if front is better than back and the heavenly motion is the best possible that nature and god can produce, this will be the explanation why it moves in the direction in which it now moves and not in the reverse. For moving towards what is more honourable, i.e. the front, is better for moving simply and unceasingly. Consequently one should answer the person who raises the difficulty why the heaven moves in this way and not in the reverse way that it is better for it to move towards the front than towards the back, since the front is that towards which natural motion takes place.

(Alexander says, 'it is also possible that something which starts from the right not always move towards the front, as is the case with those who mount, rising upward towards the back'.[284] He might also say that those who double the right[285] move towards[286] their backs, but these motions are not natural but artificial, as in the case of tumbling rather than walking.)

So the whole demonstration before us is the following: 420,1

(i) If the heaven has as its right that from which it moves in a certain direction, it will also have as its front that towards which it moves, since what moves from something also moves towards something;

(ii) so if the right and front are better than the left and back, what

> is arranged in the best way – and that is the way the heaven is arranged – will move from what is better and to what is better.

Some people consider this demonstration to be circular. For if he has proved that the heaven[287] has a right on the basis of the fact that it has a starting point of motion in a certain direction, that is, towards the front,[288] what is right would have been proved on the basis of what is front. So if he now says that the heaven has a front and a back, since it also has a right and left, the proof is circular.

But Alexander says the proof is not circular, because <Aristotle>, having proved that since the heaven is complete and has soul, it must have all the dimensions, then inferred what is right in the heaven from <its> motion; for right is that from where the starting point of its motion <comes>. But, if that from which the motion is is also right and that to which motion from the right naturally is is front, and that from which and that to which always go together, what is absurd about things which go together also being things which are recognised on the basis of one another?[289]

However, I don't think this question is worth a lot of discussion. It is better to raise the difficulty how, in inquiring why the heaven should move in one direction and not in the other and saying that it moves towards the front since this is better, he has solved the difficulty. For if in the heavenly body there were a difference between front and back of the kind that is seen in the case of animals which are generated, it would be possible to say that the motion <of the heaven> has come to be in a certain direction because this is the front. But since the heaven is homoiomerous, it is unclear whether the motion has come to be in a certain direction because this is front or whether we say this is the front because the motion has come to be in that direction. For if it moved in the other direction, rising from where it now sets and setting where it now rises in the way the planets do, it is clear that where it rose would be right, that towards which it moved would be front, and the north pole would be up, as he demonstrated in the case of the planets.[290] And the question would be why the fixed sphere moves from what is now called the west, the planets being bound to move in a contrary way from what is now the east for the reason given.[291]

But perhaps Aristotle, being more precise than we who raise difficulties, has seen the solution. For if it is necessary that both the fixed heaven and the planetary heaven have a natural front and back, and front and back are prior to motion in the case of animals, since it is necessary first to have a front and so move towards the front, and
421,1 if there is motion towards the front, it is absolutely necessary that there be a front, although, if there is a front, it is not necessary that there be motion towards the front. And in the case of the fixed heaven and the planetary heaven it is also true to say as a result that the

motion is in this direction because this is the front, but not that this is front because the motion is in this direction, which is what the person who raises the difficulty asserts.

However, no one would dispute that it is necessary that natural motion be towards the front, since the front is that towards which natural motion is directed. And I believe it is possible to verify that there is a natural front and back in the heaven in the following way. In all multiplicities which move in a coordinated way – e.g. a chorus, a procession, or an army – it is absolutely necessary that there be some order in terms of what precedes and what follows. By so much the more then did the demiurge establish the ordered multiplicity of the stars in the heaven, so that they have such and such a position and such and such configurations relative to one another and so that some things naturally precede in the motion and others naturally follow; he looks towards what is appropriate for their substance, power, and activity and fashions[292] sublunary things in accordance with this. So these things have been ordered[293] in position and precedence in the way that it was best for them to be ordered; and if they moved in the reverse way, what now precedes would necessarily follow, and Pisces and Aquarius and what follows them, but not Taurus and Gemini and their followers, would follow Aries; and after the equinox in Aries it would not be spring, as it is now, but autumn, and after Libra it would be summer;[294] and in this way the order and power of the <astrological> combinations of the sun with both the zodiac and the stars rising alongside it and their products beneath the moon would take on a different governance. For when the sun came to be in Pisces it would make the start of autumn and not the end of winter, as now. So, since, if the motions and activities of things in the heaven came to be in the other direction, it would be necessary that their powers also be different and obviously their substances as well, let the person who asks why the heaven moves in a certain direction but not in another recognise that he is not simply asking this; rather he is asking why the cosmos has been ordered in the way it is now but not in some different way. There is then no difference between putting the difficulty this way and putting it that way. So it is well that in the end Aristotle has reduced the argument to what is best.

Finally it has become clear in this passage why when he proved that there is an up and down and right and left in the heaven, he left out front and back.[295] For, as we can easily discover from the difficulty here, he postponed the discussion of front and back until this passage.

[Chapter 6]

288a13-27 After what has been said, the next thing which should be discussed concerns its motion, that it is uniform [and not non-uniform. (I say this about the first heaven and the first 422,1 motion, since in the lower <regions> the motions which are multiple have already come together into one.)

288a17 For if it moved non-uniformly, it is clear that the motions would have an acceleration, a maximum speed, and a deceleration. For every non-uniform motion involves deceleration, acceleration, and a maximum. A maximum occurs either at the beginning of motion or at its finish or in the middle; for example, perhaps the maximum for things which move naturally is the finish, for those which move unnaturally it is the beginning, and for projectiles, it is in the middle. But there is no starting point, finish, or middle of motion in a circle, since it simply has no starting point or limit or middle; for it is temporally eternal, drawn together in length, and unbroken. So, if there is no maximum speed of its motion, there will not be any non-uniformity either,] since non-uniformity is the result of deceleration and acceleration.

This is the sixth issue proposed concerning the heaven in this book:[296] concerning its motion that it is uniform and not non-uniform. He shows this because it was worthy of being handed down for hymning the praises of the divine body and because he has in fact also used it in his discussion of the shape of the heaven when he said that the revolution of the heaven is the measure of motions because it is continuous, uniform, and eternal.[297] However, these things are attributes of each of the heavenly spheres, since they are all divine and have an eternal and continuous motion. But because each of them has several motions, one its own,[298] one a revolution together with the fixed sphere, and others caused by the so-called counteractive[299] spheres, they have a certain apparent non-uniformity, even if it is not a real one. And this is the wonderful problem for the astronomers: with certain hypotheses being given by them, to prove that all things in the heaven have a circular and uniform motion and that the apparent non-uniformity in each of them (i.e. that they sometimes move faster, sometimes slower, sometimes forward, sometimes backward, and up and down, and that there are times when they rest at the same place and that the same things are seen as greater or smaller – all these things and others of the same kind) is merely apparent and not real. Such hypotheses were given both by earlier people, the associates of Eudoxus and Callippus,[300] whom Aristotle follows, namely the so-called counteractive spheres (which Aristotle explains in the *Metaphysics*)[301] and by later people who hypothesised certain eccentric <spheres> and epicycles. <Aristotle> will prove the uniformity of these things when he discusses the counteractive spheres in the *Metaphysics*. Now he says that it will be proved that the first motion is uniform, although, I believe, all the demonstrations also fit the planetary motions. But because the first motion also appears to be uniform and does not need any hypotheses, he announces now that he will argue about it.[302]

(288a17) I believe that the first argument is carried out categorically in the second figure as follows:

(i) Every non-uniform motion involves acceleration or a maximum 30
speed or deceleration;[303]
(ii) motion in a circle does not involve acceleration or a maximum speed or deceleration;
(iii) therefore, motion in a circle is not non-uniform.

He takes as clear the major premiss which says that non-uniform motion involves acceleration or a maximum speed or deceleration; for faster and slower are differentiae of motion; and the non-uniformity
of a faster motion occurs in acceleration of the motion, which becomes 35
ever more fast, whereas that of a slower motion occurs in deceleration of the motion which always progresses towards being slower; but the maximum is seen in what is fastest, just as the minimum is seen in what is slowest.

He argues for the minor premiss, which says that motion in a circle 423,1
does not involve acceleration or a maximum or deceleration, in the second figure as follows:

(i) What has a maximum speed has either a starting point or a middle or an end;
(ii) motion in a circle has neither a starting point nor a middle nor an end;
(iii) therefore, motion in a circle does not have a maximum speed;
(iv) but if it doesn't have a maximum, it doesn't have acceleration 5
or deceleration either, since the maximum is the limit of acceleration and the starting point of deceleration.

Again he proves the major premiss here, which says that what has a maximum speed has a starting point or middle or end, by a division in the following way:[304]

(i) Things which have a maximum speed in their motions are either natural bodies or bodies having soul, that is they are either simple or composite bodies;[305]
(ii) some natural bodies move naturally, some as a result of force; 10
(iii) and those which move naturally also have their maximum speed at the end of their motion both in terms of place and of time, since coming towards their appropriate wholes they move faster and when they get there they stop moving;
(iv) but those which move as a result of force and unnaturally have their maximum speed at the starting point of their motion, since

> they move faster at the starting point because the power which moves them is with them more then; accordingly when that power diminishes little by little and ceases to act, the moving object is finally moving by dint of its own inclination; for weights which are thrown or projected upward move unnaturally as long as the power of the thing exerting force is with them, but when that ceases to act they finally move down by dint of their own inclination.

He says that for projectiles the maximum speed is in the middle. What, then, are projectiles? Certainly it is clear that they are not the simple natural bodies, since these things which move either naturally or unnaturally have already been explained. Rather they are the composite <bodies> of animals,[306] which do not move up <naturally> because they have components weighing them down and do not move down <naturally> because of components of which the motion is up, but move sideways. He calls them projectiles either because of their similarity to missiles and spears which are said to be projected (since when they are discharged sideways we speak of their being projected); or he calls the bodies of animals projectiles because the souls in them 'throw' them sideways contrary to the natural motion of each of their components, since if they moved without impulse in their soul they would not move in this way but in the direction of the natural <place> of the dominant simple body in them.[307] And it is possible to learn by induction that the motions of things having soul have their maximum speed around the middle. For animals which fly or walk need to exercise and prepare their limbs for moving at the start of their motion, and, having laboured, they finally move more lazily towards the end. And this is also clear in the case of bodily exercises, since their maximum speed occurs near their middle.

Alexander, after giving this interpretation (that the maximum speed of animals is around the middle of their motion), adds another one, which he says is more plausible than this. He writes:

> He might say that for projectiles it is in the middle, not meaning the middle of their own particular motion, but the middle[308] of the space in which they move. The simple bodies such as earth and fire, move both naturally and contrary to nature; one of them has the maximum speed of both its natural and its unnatural motion at the lowest point, since a clod of earth moving naturally and one projected upward by force has its maximum speed at its lowest point; for the power of the thing projecting it from below is fresher at the bottom. However, the maximum speed of its natural motion is at the end of the motion, that of its unnatural motion at the start. Similarly, the natural motion of fire (i.e. when it moves upward) has its maximum speed on high and so does its unnatural motion when it is forcibly pushed from

> above. But the motion of projectiles, which do not move up or down but in the middle, also has its maximum speed in the middle, since the motion of each thing has its maximum speed in the region in which it moves, and projectiles move in the region between up and down (for arrows and spears also move in this region).

But perhaps Aristotle is not taking starting point, middle, and end just in terms of place, but also in terms of time and motion. For he also makes the minor premiss, which says that motion in a circle has neither a starting point nor a middle nor an end, clear, recalling it with the words 'for it is temporally eternal, drawn together in length, and unbroken', so that these things which are circular have no starting point, middle, or end, either in terms of time or in terms of motion or distance. For the words 'drawn together in length and unbroken' are also said about motion in a circle, since the preceding argument was about this; and so they are expressed in the feminine gender[309] even though they are said metaphorically <of motion> and they are truly said in the strict sense of magnitude and length, being drawn together because a thing is curved and not straight (since what is straight has a starting point, a middle, and an end), being unbroken because the thing is drawn together in such a way that it is not broken and so, there being no extrusion due to a break, there would not be any starting point or end or non-uniformity.

And Alexander himself in proceeding through these statements of Aristotle which I have set out, says:

> It is made clear that he does not mean by the middle what is between up and down since he says it is temporally eternal, drawn together in length, and unbroken. So if motion in a circle has neither starting point nor middle nor end, but what has a maximum has either a starting point or a middle or an end, then motion in a circle does not have a maximum. Consequently, if what doesn't have a maximum speed doesn't have acceleration or deceleration (since a maximum speed is always either the limit of acceleration or the starting point of deceleration), then 425,1
> circular motion has neither acceleration or deceleration. But non-uniform motion has a maximum speed and either acceleration or deceleration; therefore, circular motion is not non-uniform.

Alexander has correctly noted that even on the basis of what Aristotle says it is not necessary that everything which moves non-uniformly have an acceleration, a maximum speed, and a deceleration. For it is not true that things which have a maximum speed at the start of their motion[310] will have an acceleration. For acceleration is a path to a maximum speed and occurs before the maximum, but these things start with their maximum speed. Nor again can things which have their maximum speed at their end[311] have deceleration and a minimum speed. For the deceleration, when it is in some things, comes after the maximum speed, but the maximum speed for these things is at the end. So when Aristotle says, 'for every non-uniform motion involves deceleration, acceleration, and a maximum', he does not mean that every <such> motion involves all of them, but that the totality of motions involves all of them, and each motion, if it does not involve all of them, always has a maximum speed and either acceleration or deceleration; for in the case of things <moving> naturally there is acceleration and a maximum speed, in the case of those <moving> unnaturally there is a maximum speed and deceleration, and, as he says, in the case of projectiles there is acceleration, a maximum speed, and deceleration.

I think it is worth considering whether one couldn't deny that the motion of the planets is non-uniform on the basis of the same premisses. For it is also true of the planets that <their motion> is temporally eternal, drawn together in length, and unbroken, so that it has no starting point, middle, or end with respect to time or motion or distance. And if it doesn't have these things it doesn't have a maximum, since a maximum lies in one of them. And if it doesn't have a maximum it doesn't have acceleration or deceleration; and if it doesn't have a maximum or acceleration or deceleration, it doesn't have non-uniformity either. So, when Aristotle, in speaking about motion, said that it is uniform, and added 'I say this about the first heaven and the first motion', he did so because he was making his argument about the first heaven and wished to bear witness for the way it appears.[312]

> **288a27-b7** Furthermore, since[313] everything which moves[314] is moved by something, [it is necessary that a non-uniformity in the motion be produced either by the cause of motion or by the thing moved or by both. For if the cause of motion did not produce motion by the same power or if what is moved altered and did not remain the same or if both changed, nothing would prevent what is moved from moving non-uniformly. But none of these can happen in the case of the heaven. For it has been proved that what is moved is first and simple and does not come to be and is not destroyed and is completely unchanging, and it is much more reasonable that the cause of motion be like this. For what is first is kinetic of what is first, what is simple is kinetic of what is simple, and what does not come to be and is not destroyed is kinetic of what does not come to be and is not destroyed. So, since what is moved, which is a body, does not

> change, what moves it, which is incorporeal, would not change.] Consequently it is also impossible for the motion to be non-uniform.

This is the second argument proving that the motion of the first heaven is not non-uniform. It reaches this conclusion on the basis of two axioms. The first says that 'everything which moves is moved by something', and it has been demonstrated in Book 7 of the *Physics* that this is so.[315] The second comes after this and says that the non-uniformity of a motion comes either from the cause of motion (when it is different at different times or when it is the same but moves with a different power at different times) or from what is moved (when it is differently disposed at different times and is not always able to respond to the power of the cause of motion in the same way) or from both (when they make use of different powers). So, a non-uniform motion occurs when either the cause of motion or what is moved or both change. But no such thing can happen in the case of the heaven. For neither it nor what causes it to move nor both can change. Thus the conclusion: it is impossible that any non-uniformity occur in the case of the heaven.

The inference here is in the second figure. He has taken the major premises[316] to be clear, but he proves the minor premiss, which says that in the case of the heaven there can be no change either of the cause of motion or of what is moved or of both. He first proves it for what is moved, using what has been demonstrated already[317] about it: what is moved is the first body, it is simple, and it doesn't come to be nor is it destroyed, as has been proved, and what is of this kind is absolutely and completely unchanging. He then proves that what causes the heaven to move is unchangeable, arguing on the basis of 'the more'.[318] For if [319] the first thing moved is both simple and does not come to be and is not destroyed, and if the thing moved is a body and what causes it to move incorporeal, then by much more will what causes it to move be unchanging and unaltering. And he proved in the eighth book of the *Physics* that the first mover is completely motionless.[320] So if each of them is unchanging, it is clear that both together are as well; but a non-uniform motion would result <only> if either or both of these changed; therefore it is impossible for the circular motion to be non-uniform.

The words 'or if both changed, nothing would prevent <it> from moving non-uniformly' are an expression of caution. For it is possible that both might change at some time and there be uniform motion, if the change were proportional. That is why he says 'nothing would prevent'. Alternatively <one might say that> even if the motion were uniform because of the proportionality of the change, still such a motion would not be the same as the motion before the change, and in this sense it would again be necessary for it to be non-uniform.

He has made the comparison of the moving body with what is unmoving and incorporeal cautiously.[321] For changing applies naturally to a body rather than to an incorporeal and to a moving thing rather than to an unmoving one, since such changes attach to bodies universally.

He seems to have left out another cause of non-uniformity in a motion, the difference of the medium through which the motion occurs. This is not the same as either the cause of motion or what is moved. For if the medium is fine, the motion through it is faster, but if it is denser, the motion is slower. However, since the heavenly body is continuous with itself and has nothing outside it, it has nothing as a medium.

And clearly this argument too fits not only the first heaven but also the things which are said to wander. For each of them is moved by a cause of motion which is motionless and incorporeal, and each of them is of the primary nature, is first and simple, does not come to be and is not destroyed, and is consequently unchanging. And so much more will the things which cause them to move, since they are incorporeal and motionless, have these properties. So if in the case of these things both the cause of motion and what is moved are unchanging, it is also impossible that their circular motion be non-uniform; and if there appears to be some non-uniformity in them, it is only apparent. And that is why the astronomers, as I said previously, assume certain hypotheses and preserve the phenomena, agreeing that all heavenly things move uniformly.[322]

288b7-22 And if <the motion> became non-uniform, [it would either change as a whole and become faster at one time and again slower at another or its parts <would change>.

288b9 It is evident that its parts are not non-uniform, since if one part moved faster, another slower, a distancing of the stars would have already occurred in infinite time. But none of them is observed to change its distances.

288b12 But neither is it possible for the whole to change. For the deceleration of anything results from loss of power, but loss of power is unnatural. For in animals, every loss of power, e.g. old age and decay, is unnatural. For it may be that the entire structure of animals is composed from the sort of things which differ in their proper places, since none of their parts occupies its own region.

So, if there is nothing unnatural among primary things (since they are simple, unmixed, and in their proper region, and they have no opposite), there will not be any loss of power either, and so no acceleration or deceleration.] For if there is acceleration, there is deceleration.

In the second of the arguments proving that the motion of the heaven is uniform,[323] he assumed that everything in motion is moved by something and that, if motion in the heaven were non-uniform it would come either from what is moved or from the cause of motion or from both, and he proved briefly that a non-uniformity in the heaven cannot come from what is moved and – by much more – it cannot come from the cause of motion. In what follows[324] he divides up that argument by means of different ideas and in this third argument he proves that the motion of the heaven cannot become non-uniform from the thing in motion, and in the next he shows that it cannot become non-uniform from the cause of motion.

And he proves that it does not come to be non-uniform from what is in motion as follows:

> (i) If it did, either the whole motion, i.e. the motion of the whole body, would change and become faster at one time and again slower at another or the parts of the motion would change (he means by parts of the motion the motions of the parts of the heaven, just as he means by whole motion the motion of the whole);
> (ii) so, if it were proved that neither the whole nor its parts change their motion in such a way as to accelerate at one time and decelerate at another, the heaven would not move non-uniformly.

(288b9) He first proves by reference to the phenomena that there is no non-uniformity in the parts. For if in so many myriads of years from which a record of the position of heavenly things has been 428,1
preserved, no change in the relation of the stars to one another and no difference in their distances has been observed, just as contemporary star maps show no change relative to earlier ones, it is clear that there has been no non-uniformity in the motion of the parts <of the heaven> relative to one another in the sense that some parts moved faster, some slower. For if they had, their distances and their configurations relative to one another would certainly have changed.

(288b12) If all the parts moved faster or all moved slower at the same time, the resulting non-uniformity would not attach to the parts, but to the whole; and he demonstrates that such a non-uniformity is impossible in the following way:

> (i) Non-uniform motion comes from deceleration and acceleration;
> (ii) deceleration always results from loss of power when the proper power of a thing is unable to be sufficent for similar and uniform activity forever;
> (iii) every loss of power is unnatural.

The witness for the third premiss is that also in animals loss of power and being at a minimum are unnatural, as we see in the case of old age and sickness. And he adds the cause of loss of power in animals and obviously in plants by saying that their structure is composed from bodies which are not themselves completely in their natural condition; for it is composed from the four elements, each of which has a different natural place, which in some cases is not the same as that in which the animals and plants are. For none of the parts or elements of animals remains entirely in its[325] region, since these things are not strictly speaking in earth or in water, nor are they in pure air or fire; rather they are in the region around the earth in which the elements are mingled; and in the parts themselves of animals, earth is aggregated above and fire[326] below. So, since they are disposed in an unnatural way, they labour, and they become the cause of the loss of power in things composed from them and also of their being arranged in an unnatural way.[327] And there is a point to his now adding the explanation of loss of power and being unnatural, namely, in order to prove that since the divine body is simple and not composed from opposites and also always situated in its own region, it does not ever come to be unnatural, it does not lose power, it does not submit to deceleration and acceleration, and it does not move non-uniformly. To state matters briefly:

(i) The divine body is not constituted in an unnatural way;
(ii) what moves non-uniformly is constituted in an unnatural way;
(iii) therefore the divine body does not move[328] non-uniformly.

The inference is in the second figure.

Since loss of power is directly the cause of deceleration, he says, 'there will not be any loss of power either', and adds, 'and so no acceleration'. But since deceleration occurs after acceleration, not acceleration as a whole,[329] but the acceleration which is relative to deceleration, he adds, 'or deceleration. For if there is acceleration, there is deceleration'. For these things are and are spoken of in relation to one another, like more and less and greater and smaller.[330]

429,1 Alexander raises as a difficulty how, in things in which the maximum speed is at the starting point and there is deceleration after the maximum, as has been said for things moving in an unnatural way,[331] the deceleration can come after the acceleration, when there is no acceleration at all in this case; or how, in things in which the maximum speed is at the end, as in the case of things which move naturally, the acceleration can come after the deceleration, when there is no deceleration at all in that case. And he solves the difficulty by saying:

> Perhaps he does not say that these things come about in turns by necessity, but that they go together; for what decelerates decelerates with respect to the acceleration which is present in

> it, since if there were no acceleration present in it, it would not be capable of decelerating either. And similarly, what accelerates accelerates insofar as it participates in deceleration, since, if there were nothing decelerating in it, there would be no place for acceleration.

And what he[332] says is true, since acceleration involves what follows being faster than what precedes, and deceleration the reverse. It is perhaps also possible to say in the case of things in which deceleration comes after the maximum speed, as when deceleration is spoken of relative to a maximum speed reached after acceleration, and in the case of things where the acceleration is before the maximum speed, as in the case of things <moving> naturally, that the <motion> which proceeds to a maximum speed is said to accelerate with respect to an original disposition to decelerate.

Perhaps, Alexander says, it is also true that decelerations which come about as a result of choice have the same cause, i.e. loss of power, since when we fear that our power is being used up we employ deceleration.

> **288b22-7** Furthermore, it is also unreasonable that the cause of motion is unable to act for an infinite time [and then able to act for another infinite time. For nothing is observed to be in an unnatural way for an infinite time (but loss of power is unnatural); nor are what is natural and what is unnatural, or in general what is possible and what is impossible, observed to be for equal times. But it is necessary that if the motion decelerates] it decelerates for an infinite time.[333]

Having proved from consideration of what is moved that the heaven cannot have a non-uniform motion, he now gives an argument from consideration of the cause of motion. It seems that he carries it out from a division of the following kind:

> If the heaven moved non-uniformly its motion would either accelerate for an infinite time and decelerate for an infinite time or just decelerate or accelerate *ad infinitum* or at one time accelerate and at another decelerate during certain alternating cycles.

And he proves the first <alternative is impossible> by means of the second hypothetical <mode>:[334]

> (i) If the motion of the heaven decelerated for an infinite time and accelerated for an infinite time, then, since its acceleration and deceleration would result both from the loss of power in the cause

of motion and from its power, the cause of motion would lose power for an infinite time and gain it for an infinite time;
(ii) so if loss of power is unnatural, the cause of motion would be in an unnatural state for an infinite time;
(iii) but this is impossible, 'for nothing is observed to be in an unnatural way for an infinite time', because what is unnatural is a turning away from and a byproduct of what is natural;
35 (iv) therefore, the antecedent from which this followed is imposs-
430,1 ible, but that was that the motion of the heaven decelerates for an
infinite time and accelerates for an infinite time.

And he infers another conclusion from the same premisses:

(i) If the cause of motion is possible, that is, natural, for an infinite
time and impossible, that is, unnatural, for an infinite time, it
5 would be unnatural and natural, which is the same as to say
possible and impossible, for an equal time;
(ii) but it is absurd that what is unnatural last for the same amount of time as what is natural, or what is impossible for the same amount of time as what is possible, since the natural and the possible are primary, and what is unnatural and what is impossible are miscarriages of them;
(iii) consequently the antecedent from which this followed, namely
10 that the motion of the heaven decelerates for an infinite time and
accelerates infinitely, is in turn absurd and impossible.

Alexander says:

He uses acceleration for an infinite time and infinite deceleration, and he has not demonstrated these things; rather he has
not mentioned that the acceleration is also complete,[335] and only
15 said this much about the deceleration: 'but it is necessary that
if the motion decelerates it decelerate for an infinite time'. For if the first mover is the thing which causes this motion which loses power eternally, what will restore its loss of power? So, if it always causes motion, it always does so because of its loss of power. But he has left out that the acceleration is to infinity, since it is clear that if the loss of power, which is unnatural, is
to infinity, the power, which is natural, will be even more to
20 infinity; for it is necessary that what is natural not last less long
than what it unnatural, when they do not last equally long either.[336]

And in saying these things, Alexander correctly recognises that Aristotle is not now proposing to prove that if <the heaven> moved non-uniformly, decelerating and accelerating, it would decelerate *ad infinitum* and accelerate *ad infinitum*, but rather – <in the context of> proposing to prove that the motion of the heaven is uniform – he is eliminating all ways of being non-uniform, considering them on the basis of a division, as I explained before.[337]

Alexander says that the text is rather unclear because <Aristotle> has not set out the division first. And from his taking it as based on division Alexander of Aphrodisias himself seems to be the first person to pay attention to the demonstration. For he ascribes to Alexander of Aigai[338] the interpretation just given according to which deceleration *ad infinitum* is necessary because there is nothing to restore the power of the first mover and make good its loss of power:

> I heard <what> Herminus <said>, in agreement with what was also put forward in the lectures of Aspasius:[339]
>
> 'If in the case of the divine body there were deceleration, there would have been acceleration previously, since every deceleration is attached to an acceleration. However, if there were an acceleration before the deceleration, it would have been *ad infinitum*. For it is not possible to provide a reasonable explanation why, if the motion is infinite, the acceleration continued as 431,1
> long as it did and did not stop earlier. However, if there is also a deceleration, it will in turn be *ad infinitum*. For if the body moved *ad infinitum*, it would not be possible to say what the explanation is of its overcoming the deceleration or to give an explanation for why it decelerates as long as it does. So the deceleration is also *ad infinitum*. For the deceleration is a deceleration of an acceleration, and if the acceleration is infinite, so is the deceleration. For why should the deceleration be of a part of the acceleration and not of all of it? However, the deceleration comes from loss of power, but that is unnatural, and what is unnatural cannot be equal to what is natural, since what is unnatural is a path to destruction. Consequently there will not be deceleration. And if there is no deceleration, there won't be any acceleration either.'

This, then, is what Herminus said in agreement with Aspasius. And it is clear that these people too did not understand that deceleration *ad infinitum* and acceleration *ad infinitum* are assumed by reference to the necessity of the division, but they think that they are taken as resulting necessarily if, because there is deceleration and acceleration, the motion is non-uniform. This is clear, but it is also clear that what is refuted by Aristotle is only the idea that both the acceleration and the deceleration might be absolutely infinite. The next parts of the division make clear that <infinite deceleration and acceleration> are not taken in the other way <by Aristotle>. In them he says, 'however, it is not possible either that the motion accelerate forever nor again that it decelerate forever',[340] and again, 'there remains <the possibility> of saying that fastness and slowness alternate in the motion'.[341] In these passages he refutes the other parts of the division, which do not hypothesise that acceleration and deceleration are both infinite, but that either there is always acceleration or there is always deceleration or they alternate.

It is worth inquiring why Aristotle has added the words 'but it is necessary that if the motion decelerates it decelerate for an infinite time'. For he caused the previous interpreters to understand this word 'necessary' not as expressed on the basis of a division and to seek the reason for the necessity. Alexander of Aphrodisias, who recognises this, says:

> One should supply the words 'after the infinite acceleration' at the end of the sentence. For if the motion decelerates after an acceleration for an infinite time, and there is not only acceleration but also deceleration, it will decelerate for an infinite time. This was also hypothesised in a different way,[342] but nevertheless he also proves that it follows necessarily from the preceding acceleration *ad infinitum*. For if the deceleration starts to occur because of a loss of power in the cause of motion, it will then continue *ad infinitum*. For what will restore the loss of power of the cause of motion, if the motion is eternal and the deceleration increases forever? The deceleration will be *ad infinitum* because of this and because the deceleration is infinite, if the acceleration is. And in this way what is unnatural and what is natural will be for an equal time, which is absurd.

But perhaps it is not necessary to supply the words 'after the infinite acceleration'. For Aristotle, using a division, is eliminating the ways of being non-uniform; he sets out the one which says that there is acceleration for an infinite time and in turn deceleration for another infinite time; and he refutes it in two ways: on the basis of the fact that nothing unnatural exists for an infinite time, and on the basis of the fact that nothing is for equal amounts of time unnatural and natural. And he remarks that the assertion that there is deceleration for an infinite time follows from the assertion that the motion of the heaven decelerates at all not only on the hypothesis <that there is an infinite acceleration>, but it also follows necessarily if the motion is assumed to be eternal because there is nothing to restore the loss of power of the first cause of motion, since it transcends everything. It does not follow because an infinite acceleration precedes the deceleration; for, even if this be true, Aristotle is obviously here not taking it as necessary but as a hypothesis, recognising in the case of deceleration only that it is necessary that if the motion decelerates at all it decelerate for an infinite time.

288b27-30 However, it is not possible either that the motion accelerate forever [nor again that it decelerate forever. For then the motion would be infinite and indeterminate, but we say that every motion] is from something and into something and determinate.

He has proved that, if it is hypothesised that the motion of the heaven is non-uniform in the sense that both its deceleration and its acceleration are infinite, two absurd consequences follow: the assertion that what is unnatural is eternal, and that what is unnatural lasts for the same amount of time as what is natural. He now sets out the second part of the division, which says that the motion is non-uniform because one or the other is *ad infinitum* – that it either accelerates forever or decelerates forever; and he proves that this is also impossible as follows:

(i) Every motion is from something and into something and is determined by its whence and whither;
(ii) but a motion which accelerates forever or decelerates forever is not determined and doesn't have a whence and whither;
(iii) therefore, no motion either accelerates forever or decelerates forever.

That what accelerates forever is indeterminate and has no whither is clear from the following:

(i) In the case of things which move naturally acceleration occurs as the thing gets ever nearer to the place to which it moves;
(ii) in the case of what accelerates *ad infinitum*, there would be no place to which it moves (since the place to which what accelerates in its motion moves will always be other than its previous place, and, because it undergoes this *ad infinitum* it will pass into a different place *ad infinitum*);
(iii) consequently, a motion which accelerates forever will not have a place to which it moves, it will not have a whence and whither, and it will not be determined.[343]

However, it is possible for something to move *ad infinitum* in the same <position> because it moves with the same motion again and again, as we say in the case of circular motion, but it is not possible for something to accelerate *ad infinitum*. For the motion will not remain the same; rather the present motion will be different from its predecessor and not similar to it, since there will always be an addition to it. And if there were something which accelerated in a circle in this way, it would strive either to fill out a certain number of revolutions or to proceed to the limit of the speed of motion, but 433,1 neither of these things occurs in the case of what moves *ad infinitum* with an acceleration of this kind.

The same argument applies to the case of deceleration *ad infinitum*; for in the case of things of this kind there is no whither and no place to which the motion goes since the maximum speed is always near the place to which the motion goes, but things that decelerate forever always get further from the maximum speed and from the place towards which they move. So, if every natural motion has a whence and whither, and no motion which decelerates *ad infinitum* has a whence and whither, then there is no motion which decelerates *ad infinitum*. And since the construction of the arguments is the same in the case of acceleration *ad infinitum* and deceleration *ad infinitum*, Aristotle has carried it out once for both cases.

> **288b30-289a4** Furthermore, if someone were to assume that there is some least time [and it is not possible for the heaven to move in less (for just as it is not possible to play the cithara or walk in any time whatsoever, but for every action there is a determinate least time which is not overcome, so too it is not possible for the heaven to move in any time whatsoever) – if, then, this is true – there will not be an acceleration of motion forever.[344] And if there isn't acceleration, there won't be deceleration either, since both together and each one singly are similar, if it accelerates at the same[345] speed or a greater one,] and for an infinite time.

This is another argument establishing with great clarity the same thing: that it is not possible for the motion of the heaven to be non-uniform in the sense of accelerating forever or decelerating forever. He first assumes an axiom according to which 'there is some least time and it is not possible for the heaven to move in less'. He first confirms this axiom by induction, using the fact that there is a certain least time for every action and everything which moves, and it is not possible for the motion to occur in less time, for example, running a stade, or playing a tune on the cithara, or singing a song. But if this is so, there will also be some least time for the revolution of the heaven from a point and back to it, and it is not possible for it to revolve in less. However, it is not universally true that there is a least time because everything continuous is divisible *ad infinitum*. But, relatively speaking, there is a least time for some things, as there is also a least time for motions. This being so, if what accelerates moves the same distance in less time, then, if something were to accelerate forever, it would always reduce the time and move a distance in proportion to its acceleration. But if this is so, then there will sometime be a time in which it[346] will move which is less than the least time in which it moves. But this is impossible; therefore what this followed from is also impossible, namely that what moves accelerates forever.

Having proved in this way that there is no acceleration *ad infinitum*, he adds that if there isn't acceleration, there won't be deceleration either. For whatever is the case with one of the two is the case with both, because they are and are spoken of relative to one another, like more and less. For an infinite deceleration attaches to an infinite acceleration, and vice versa. For nothing decelerates unless it has accelerated, and it decelerates just as much as it has accelerated. Consequently, if it has been proved that there is no acceleration to infinity, it has been proved there isn't deceleration to infinity either.[347]

Since acceleration occurs because of addition, deceleration because 434,1
of subtraction, but not every addition produces an acceleration in a simple way so that there is also deceleration, after saying that both together and each one singly are similar, he adds an explanation of what manner of addition brings about an acceleration relative to which it is necessary that a deceleration be produced and for which it is true that both together and each one singly are similar. For this is true if either the additions attached to the thing said to accelerate are equal to one another or a successor addition is greater than its predecessor, since only then is there just acceleration and acceleration in the strict sense. But if the additions proceed in the direction of getting less – if, say, first a speed of ten breadths of a finger[348] were added and then nine and so on – first of all such a <motion> is not just acceleration but also deceleration (and so there is not some other deceleration of this <acceleration>) and furthermore it is not possible for this acceleration to go on for an infinite time, since it will <eventually> be spent because what is added will always be diminishing. It is because of this that, having said 'since both together and each one singly are similar', he adds 'if it accelerates at the same speed or a greater one, and for an infinite time'. And in fact, if the subtractions are equal to one another or one is greater, another less, the deceleration remains a deceleration, but in that case the deceleration is proportional to the acceleration, if the subtractions are proportional to the additions.[349]

> **289a4-8** There remains <the possibility> of saying that [faster and slower alternate in the motion. But this is completely unreasonable and similar to a work of fiction. Furthermore, if this had been the case, it is more reasonable that it would not have escaped notice,] since things that are placed alongside of each other are easier to perceive.

What comes next is, in fact, the third part of the division; it says that the motion is non-uniform because faster and slower are present in it in alternation. And he refutes this hypothesis first on the basis of the fact that it is not reasonable to say that there is a cause of the heaven moving, say, faster today and slower tomorrow because, among writings, works of fiction assert what is not the case and they are not able to give a reasonable explanation for what they say. He adds a second refutation to what he says, namely that if this happened it wouldn't escape notice. For differences and dissimilarities and opposites, whether they are perceptible or intelligible, can be distinguished more easily when they are positioned next to one another.

> **289a8-10** Let us[350] say this much about the fact that the heaven is only one, [that it does not come to be and is eternal, and further that it moves uniformly.]

He adds this conclusion, which includes both what was said in the first book and in this one about the heaven. For it has been proved 435,1 that it is simple, that it moves with a simple, circular, and uniform motion, and that it is only one. It has not been shown that it is one sphere (since he knows that the planetary is different from the fixed and he knows the division of the planetary into several <spheres>), but that there is one cosmos with a system of eight spheres and not several heavens which fill out more than one cosmos. It has also been shown that it does not come to be, not in the sense that it is not caused, but in the sense that it is eternal and has not come to be at a particular time; nor is it at one time and not at a later one. It is reasonable for him to have set out this conclusion now when he has filled out what is to be said about the heavenly body and is turning the discussion to the study of the stars.

[Chapter 7]

> **289a11-19** We should next speak about what are called the stars[351] [and say what they are composed from, what kind of shapes they have, and what their motions are.
>
> 289a13 It is most reasonable to say and it follows from what we have said that each star is composed of the body in which it has its motion, since we said that there is something which is of such a nature as to move in a circle. For we are speaking in the same way as those who say that the stars are fiery because they say the upper body is fire and that it is reasonable that each thing be composed of that in which it is.]

This is the seventh topic proposed for discussion in this book.[352] After the study of the spheres as wholes and their inherent parts, which are studied in terms of their spatial relations, right and left, up and down, front and back, and after the manifold issues concerning their motions, he turns now to the study of the stars which are contained in them and proposes to investigate in a general way all the stars, inquiring what substance and what kind of shape they have, and in what way they move; later[353] he will work through some of the difficulties concerning the difference between the wandering stars and the fixed ones.

(289a13) And so he says first about their substance that the stars are of the same substance and body as the spheres in which they are contained and carried around. He says that this follows from the things he has said because, on the basis of the two simple and primary motions, motion in a straight line and motion in a circle, he hypothesised that there are two kinds of primary bodies, one moving in a circle, everything heavenly, the other progressing in a straight line, sublunary things.[354] So, the stars, since they are in the heaven and move with the heavenly motion, would be of the heavenly substance; and even if our eyes tell us that the body of stars is different from the heavenly body, it is not thereby necessary that they have a different substance. He confirms this by referring to those who say that the stars are fiery in appearance because they also assume that the whole heaven is fiery; but perhaps those who assume this also assume that the heaven is fiery because the stars are.

The divine Plato,[355] who says that the whole cosmos is composed out of the four elements, thinks, for the reasons mentioned previously, that the heaven is composed of the highest forms of the elements. And, just as in the case of bodies here, which are also composed of the four elements and are last, the last form of the four elements, the earthy one, dominates, so too among divine bodies, which are composed of the highest forms of the elements, the highest form of the first element, fire, dominates. And so he writes in the *Timaeus*:

> <The demiurge> fashioned the greatest *idea* of the divine <kind of living thing> from fire[356] so that it would be the most beautiful and the brightest to see.[357]

Alexander sensibly raises this difficulty:[358]

> Why, if the so-called fifth substance of the body which moves in a circle is simple, is there such a great apparent difference between the body of the stars and the heavenly body? If they differ entirely because of denseness and rareness or because of colour or some other properties of this kind, how can they be called simple or not subject to affection, since affections occur because of these differences and these differences are affections?

And he says:

Just as was also said previously, there is nothing anomalous about there being such differences and qualities in the heavenly body, if they are inherent and do not occur because of being affected. For <the heavenly bodies> are not free of affection without qualification; rather they do not suffer 'any of the difficulties to which mortals are subject',[359] and they are this way because no opposite into which they could change and be destroyed attaches to their substance and the form with respect to which they are. There is, in fact, nothing anomalous about some opposition attaching to them and their being related to certain things which do not share at all in their substance. For, in fact, in the case of bodies which are destroyed and come to be there are some things to which they are opposite, as in the case of snow there is white and cold;[360] for, as long as it is snow, white is inseparable from it, and snow does not change in colour just because there is some opposite of white. For as bodies the <heavenly bodies> are simple, although none of them is absolutely simple if they are composed of substratum and form.

This is what Alexander says in these words: that some qualities which are in opposition to one another are present in the heavenly <bodies> in the case of colour and of size (and perhaps also of shape; for there are rectilinear shapes among the heavenly configurations, e.g. in the case of the constellation Triangle, and curvilinear ones, e.g. in the case of the Crown),[361] and that there – if one accepts the view of Alexander – these oppositions are separate from the difficulties to which mortals are subject and affections which produce destruction.[362]

But it seems to me worth considering <Alexander's> assertion that, even if there are opposites to the white and cold in snow, snow does not change in colour or qualities because of this. For if he says that if the snow endures it does not change with respect to these things, what he says is true, but it is not, I think, relevant to the issue at hand. For snow does not change while something else remains just what it is, but rather the substratum remains and, taking on differences from other things, is said to change. But even if <it is true> that it doesn't change just because of the opposition, how is <Alexander's> statement true, since until now we have always said that coming to be and destruction lie in opposites, because things that come to be come to be from opposites and things that are destroyed are destroyed into opposites. And when snow (which is compacted cloud that has
437,1 already changed into water but still has much which is airy and sort of foamy and as a result has become white, since coldness has aggregated it into a form of this kind) draws near to heat, it is divided again into the things from which it was composed, and the white departs with the part of it which is heated and is turned into air, and

what is made liquid is turned into water. And if heat were not the opposite of coldness, liquefaction and aggregation the opposites of coagulation and division, snow would not be dissolved at that time.

So perhaps it is true that what comes to be and is destroyed has its existence in opposites, and where there is no opposition it is impossible for there to be coming to be and destruction. However, it is not in fact also true that where there is opposition, it is necessary that there be coming to be and destruction. Rather, this is true only in the case of those opposites which are of such a nature as to come to be in the same substratum but are unable to exist together with one another; for these are of such a nature that, contending with one another for existence and fighting for existence, they destroy one another. But because the *logoi*[363] in the knowledge of opposites, of white and black and of hot and cold, are unextended and exist in unextended things and are of such a nature as to exist together with one another, they not only do not destroy one another, they also can bring one another into existence.[364] And the white in knowledge does not fight with the black in matter, nor do these two destroy one another, nor do they come to be from one another, because they are not of such a nature as to be present in the same substratum. And so in the same way the denseness in the heaven, for example, and the rareness here do not fight with one another, nor do they change into one another, since, being foreign to one another, they are not of such a nature as to come to be in the same substratum. Nor does the denseness in the heaven fight with the rareness there, nor does standing still there fight with motion there, because their substrata are different in nature, the poles, for example, being of such a nature as to stand still, the equator of such a nature as to be in motion. And, for example, the sun and the stars, which are thought to be dense, and the bodies in the heaven which resemble rare things,[365] do not change into one another because the nature of the substratum is always such as to be related to one and the same form and because of the unchanging substance of the form, which results directly from an unmoving cause.

And what I have said does not differ from what Aristotle has proved in the first book about the fact that the heaven does not come to be and is not destroyed. For he says there that everything which comes to be comes to be from an opposite and some substratum and is likewise destroyed into an opposite by the action of an opposite, there being some substratum, as was said in our first discussions.[366] And, hypothesising these things on the basis of what had been proved in the first book of the *Physics*, he next demonstrated[367] in addition that there is no opposite of the heavenly body into which it could 438,1
change with a substratum remaining the same. For if there were a motion opposite to its present motion with which it could move naturally, that would be the thing from which it changes and into which it changes. So, having proved that there is no motion opposite

to the circular motion with which the heaven moves, he inferred that there isn't any body which is opposite to the heavenly body, from which and into which it changes. Consequently, from all these things it is clear that coming to be and destruction always lie in opposites, and where there are no opposites there is no coming to be and destruction, but it is not the case that where there are opposites it is always necessary that there be coming to be and destruction unless there is also a substratum common to the opposites and this is extended and divisible with the consequence that it is not possible for opposites to exist together in it.

But the <original> difficulty raised by Alexander still seems to me to remain. For even if having opposite qualities does not prevent the stars from not coming to be and being destroyed, how can it be thought to be true that they are of the same substance as the heavenly bodies in which they move, when the differences in colours and structures and obviously in powers and activities are so great? Perhaps Aristotle himself solves this difficulty when he adds, 'since we said that there is something which is of such a nature as to move in a circle'. For, just as the sublunary simple bodies share the genus 'progressing in a straight line' in common, and they come to be and are destroyed and change into one another insofar as they do progress in a straight line, although their proximate forms differ from one another to the great extent which not only our reasoning but also our perception indicates, so too the heavenly bodies share moving in a circle in common and consequently also share not coming to be or being destroyed, but with respect to their proximate forms nothing prevents their having great difference from one another.

But we should move on to the next things said by Aristotle.

289a19-35 Heat and light come from <the stars> [because air is chafed by their motion. For motion is of a nature to set even wood, stone, and iron on fire; so it is more reasonable that it does the same to what is closer to fire, and air is closer to fire. The case is like that of moving missiles; for these are themselves set on fire to such a degree that lead balls melt; and since these[368] are set on fire, it is necessary that the air encircling them undergo the same thing. So these <missiles> themselves are heated up because they move in air which becomes fire because of the impact of motion.

Each of the upper things is moved in a sphere; so they are not set on fire, but, since the sphere is in motion, it is necessary that the air which is under the sphere of the body moving in a circle be heated up and especially there where the sun is fastened. That is why heat results when the sun gets nearer or rises up and is over us.

Let us say this much about the fact that the stars are not fiery and do not move in fire.]

Having referred to those who say that the stars are fiery because the heaven is fiery, and adopted this much of the doctrine into his own view, namely their doctrine that the stars have the same substance as the heaven, but disagreeing about their being fiery, he eliminates this claim by refuting what seems plausible in their doctrine: if things here heat and what heats is fire, how, he asks, can it be that heavenly 439,1
things are not fiery? And his conclusion, 'let us say this much about the fact that the stars are not fiery and do not move in fire', makes clear that his discussion is aimed at this issue. He makes plausible that, even if things here heat <other things>, it is not necessary for them to be fiery to heat, but their motion is sufficient for this, by reference to the fact that very solid things, wood, stone, and iron, are set on fire because of motion. And he sets out the clear example of missiles moving through air. For these are set on fire to such a degree that even lead balls which contain iron inside wood[369] melt and the air surrounding them is set on fire along with them. So, just as the air lying alongside is also set on fire by the motion of what is set on fire, so too, since the *hupekkauma*[370] is moved directly by the motion of the heaven and is set on fire <by it>, it is reasonable that the air which is closer to the fire be heated.

Alexander understands <what Aristotle here calls> air to be the *hupekkauma*,[371] which, he says, Aristotle elsewhere calls the element fire.[372] But how can <Aristotle> say that the air is closer to fire, if he considers the *hupekkauma* to be air? And how could he preserve the analogy with a missile if the missile which sets the air on fire were the analogue of fire? Alexander's interpretation is that the missile is heated by the chafed air, but that the air is not warmed by the missile even though Aristotle says that the missiles themselves are set on fire by their motion to such a degree 'that lead balls melt; and since these are set on fire, it is necessary that the air encircling them undergo the same thing'. Perhaps Alexander based his supposition on what is said next. He says,

> Things here which move through the air are themselves set on fire by the impact and chafing involved in the motion itself, whereas each of the stars themselves which are moved in the heavenly sphere assigned to it is not set on fire, since they are not of such a nature as to be set on fire.[373] But air is. (Here <Aristotle> calls the whole, including the *hupekkauma*, air.) So it is clear that it is necessary that this whole, which is under the sphere of the body which moves in a circle, be heated up, since the sphere moves and causes the whole to move along with it. For not only is the *hupekkauma* moved along with the heaven,

but so also is the air which is near to it and which extends practically as far as the air[374] which is settled on the earth and contained by the mountains.

Everyone says this, and it would be possible to raise the difficulty why, when the heaven and the stars in it move in the same way at night and in the day and in winter and in summer and the underlying air is always moved along with it as a unit in the same way, nevertheless the days are so much hotter than the nights and the summers than the winters.[375] So, to dissolve this objection he has added that the revolving heaven heats the underlying air 'especially there where the sun is fastened', and this is why more heat results in summer when the sun gets nearer to our zenith and in the day when the sun 'rises up and is over us'.[376]

And Alexander says that it is not unreasonable that the heat from the motion would be more and greater in that part of the underlying body which revolves as a unit,[377] when the part is against a great body which is compressed and not very distant,[378] since the contact with the motion in this part in a way involves more resistance because of the denseness and the size of the body.[379] So as a result the revolution of the heaven most heats the air in that part of it against where the sun is fastened.

But it still seems to me that the difficulty remains. Why does the sun, which is practically equidistant from us everywhere (since the earth has to the sphere of the sun practically the relation of a point and centre – parallax being very small), heat the air around us more in summer than in winter, although it is practically no less distant from this place in summer? And why does it heat more when it is in mid-heaven than when it is rising or setting, since it is at the same distance from what lies under it? And why, even if the body of the sun is large and dense, does the contact with respect to the motion of the heavenly body with the sublunary in the region of that part involve more resistance, when there are so many heavenly bodies between <the sun and the sublunary realm>?

However, Alexander has dissolved this objection previously[380] and he does so now by saying that many affections are produced in many things through certain intermediaries which are not affected in the same way, but transmit the affection to the things which are affected, just as kindling wood is set on fire by the sun but the intermediaries <between it and the sun> are not set on fire. Previously he gave as witness the affection produced in humans by the sea creature, the sting ray, through the intermediary of nets which are not themselves numbed.

Perhaps there is nothing surprising if certain qualities are communicated in this way and if in relation to the transmission the intermediaries are like certain receptacles without soul. But I do not think it is possible to make up a story that there is corporeal pushing and resistance of this kind through a certain intermediary, which is itself not pushed and does not offer resistance. Rather it would be necessary that a dense heavenly body under the sun be itself pushed by the part <of the sphere> near the sun and in this way transmit the affection into the sublunary realm. So if what is in heaven does not admit such affections, it is necessary to produce a solution of the difficulties in another way, and this is so most of all because Aristotle has not invoked density and resistance to account for the sun heating things here but <just> the fact that it gets nearer to us and comes to be over us. And so one should say that corporeal[381] rays emitted from the solar body move through the heavenly body to sublunary things and that, being immaterial, they traverse the heavenly body, as an immaterial thing, in an immaterial way without impediment, but that they no longer progress through the sublunary realm in this way because it involves matter. For <to make a comparison> certain souls receive the incorporeal radiations of heavenly bodies because of their suitability to do so, but others do not receive them because their sort of suitability is an impediment, and this is the way it is with sublunary things. But the radiations pass through air via its pores, and they are reflected back from solid objects at equal angles.[382] So, when the air is taken hold of by them and moves along with them and is chafed in itself, it is heated up because of the compression, and this happens most of all in those regions in which the rays are reflected back at themselves and compress the intermediate air. And as a result the summer solstice is hotter, because, when it occurs, the sun does not simply come closer to us,[383] since, as I have said, it is always practically equidistant from us, but, because it gets nearer to our zenith, it makes the rays, which are mostly at the meridian and near it, be reflected back upon themselves and it causes the air which receives them to be compressed and ground together. But when the sun is in winter or towards the east or west during the day, the rays which are sent in the direction of our location are no longer reflected back on themselves in the same way, but rather they are more open[384] and do not compress the air together in the same way.

It is because of this that mountains are colder than plains; for the rays are in closer contact when they are near the <point of> reflection itself and <then> they compress the air together, but, when they rise up, they stand further apart with the result that higher air is colder. On the other hand, if the setting on fire by the motion started from above higher things should be hotter because they are closer to the things in motion. So when Aristotle says that heat results when the sun gets nearer or rises up and is over us, he means by 'getting nearer' being at our zenith and by 'rising up' being above earth and not below it; and 'being above us' is being at our meridian. For when these three come together, that makes things here to be most heated by the sun.

And these things have been accounted for corporeally and in a physical way. And let it also be said that, just as <heavenly things> produce health and sickness, wealth and poverty, and different sublunary arts
442,1 for different people by their own motions and configurations, so too they produce qualitative differences. For if heavenly things not only heat but also cool, it is necessary to assign responsibility not just to their motions but, prior to them, their particular features.

Alexander notes that <Aristotle> proposed to speak about how both heat and light come from the stars when they are not fiery, but (as some people have thought) he has spoken only about heat because illumination of the transparent follows from heat and setting on fire. However, this is not the way things are because not only fire illuminates, but so do certain other things, namely divine things themselves. Accordingly he will say in the second book of *De Anima* that light results from the action in the transparent of either fire or the divine body.[385] Consequently the stars themselves can also produce illumination even if they can't heat, as he will say elsewhere.[386] And that is why he now passes over the discussion of this topic.

I have set out these statements of Alexander, which are serviceable in other ways, because I am surprised that he says that here <Aristotle> does not speak about how light comes from the stars but only about how heat does, even though he recognises that <Aristotle> says, 'heat and light come from <the stars> because air is chafed by their motion'. So if he says nothing about light, he doesn't say anything about heat either, since he doesn't say anything else about heat than that it comes 'because air is chafed by their motion'. Everything that is said in the sequel has been taken up to give believability to this.[387]

Let us also add these remarks of Alexander. For when Aristotle says that what is under heavenly things themselves are heated by rubbing,[388] he asks:

> How can this be true if <heavenly things> are not tangible? Accordingly, I inquired whether – since everything tangible involves some tangible opposition and primary tangibles involve the primary oppositions, and he proved in the second book of *On Coming to be and Destruction*[389] that the primary tangible oppositions are hotness/coldness and dryness/moistness – if what moves in a circle is characterised by these oppositions or by others. For if it is characterised by some of these, it is characterised by some of the four.[390] However, if it is characterised by others, first of all it would seem anomalous that the first of bodies is not characterised by the first oppositions; and then if it were characterised by some tangible opposition, it would not be eternal, since there would be something opposite to it. Perhaps, then, one should say that it is tangible and it too is

> characterised by a tangible quality, but not by an opposition. For things that come to be and are destroyed are characterised by tangible oppositions because they involve matter and necessar- 443,1 ily change into one another, but what moves in a circle is not of this kind, but rather is characterised by a tangible quality, namely circular motion, to which there is no opposite.[391] For if heaviness and lightness are tangible qualities, as he says in enumerating the tangible oppositions,[392] the circular motion, which is the form of that body, will also be tangible. And heaviness and lightness are not forms of bodies which come to be, since heaviness and lightness are not active or passive, since they are not spoken of as acting or being acted on by something else, as he says.[393] But things which come to be act on and are acted upon by one another. And the circular motion is active, but it is not acted upon because it has no opposite by which it could be acted on, since things acted on by it are not acted on as opposite <to it>. It is reasonable that this body has a form, and it would also be rare and spoken of as open-textured, to which the opposite would be dense, and it would admit more and less, and the spheres would be more open-textured, the stars less, since these seem to be more compressed. This compression does not change them in substance because this kind of rareness and density follows from the circular motion, and that is unchanging. And their colours, if some such things belong to them, will belong to them as external accidents and consequences of that quality.[394]

This is what Alexander says. And I think that a person could object to many of the things he says, and first of all to his assertion that heavenly things heat sublunary things by rubbing, just as they are chafed by sublunary things. For I do not think that Aristotle says this, but rather that the air is moved along with the motion of the heaven and is chafed by itself, since it is not reasonable that the heavenly body would offer resistance in such a way as to rub what is near to it. But if things were this way, the parts of the air near to the heaven would be hotter, but now on the contrary things near the earth are observed to be hotter than things higher up.[395]

Second, one might object to the assertion that everything tangible is characterised in terms of the four qualities. For the heavenly spheres do touch each other, since there is nothing in between to keep them separate. But if what is in heaven transcends the four elements, it will not have tangibility because of these four differentiae. But if it is tangible because of what is earthy and visible because of what is fiery, as Plato says,[396] it is certainly not because of what is earthy and fiery among things that come to be and are destroyed and change into one another but because of the highest forms of these, which exist in a prior way in harmony with one another and therefore do not harm the special character of the heaven. For it was also said previously[397] 444,1 that it is not necessary that everything compounded of opposites come to be and be destroyed, unless the opposites cannot exist together with one another and their substratum is of such a nature as to receive each of them in turn, as Alexander himself also thinks.

Third, how can he call the circular motion a tangible quality and not think that it has an opposite, since all qualities which are different are divided out of the genus by antithesis?

Fourth, how can he envisage that rareness and density, which are opposites, are present in heavenly things, if he considers every opposition to be a cause of coming to be and destruction? And how can he say that they follow from the circular motion if there is also rareness and density in sublunary things?

Fifth, how, if he accepts that there are colours and such external accidents among heavenly things, is he not constrained to say that these things occur as a result of being affected with the result that, even if they do not come to be and are not destroyed, there are always causes of their passive alteration, which also fits the difficulty to which mortals are subject?[398] However, I think that it is extremely absurd to speak about external accidents in the case of those things, when everything in them has substantiating and form-giving power.

[Chapter 8]

289b1-27 Since both the stars and the whole heaven are observed to change position, [it is necessary that the change occur either with both being at rest or <both> moving or one at rest, the other moving.

Now it is impossible that both be at rest, if the earth is at rest, since the <observed> phenomena would not occur. Let it be assumed that the earth is at rest. Then either both move or one moves and the other is at rest.

289b7 If both were moving, it would be unreasonable that the speeds of the stars and the circles be the same, since, then, each <star> would have the same speed as the circle on which it is observed, since it is observed that they are restored again[399] to the same position at the same time as their circles. The result is that the star will have gone through its circle at the same time as its circle has gone through its motion through its circumference.

289b13 But it is not reasonable that the speeds of the stars and the sizes of their circles have the same ratio, although there is nothing absurd in the circles having their speeds proportional to their sizes; rather it is necessary;[400] but that each of the stars in them <have their speeds proportional to the magnitudes of

> the circles> is entirely unreasonable. For if it is necessary that the star moving in the greater circle be faster, it is clear that if the stars were displaced into one another's circles, one would be faster, the other slower (and in this way they would not have their own motion, but they would be carried by the circles); and if it just fell out this way spontaneously, it would still not be reasonable that it happened in such a way that in every case the circle was greater and at the same time the motion of the star in it was faster; for there is nothing absurd about one or two turning out this way, but that they all would do it in the same way resembles a work of fiction. And at the same time there is no chance in natural things,] and what holds everywhere and in all cases is not a matter of chance.[401]

This is the second issue concerning all the stars, both the fixed stars and the planets, the inquiry whether their apparent motion and change of place occurs because they move on their own or whether it results in some other way. He carries out the inquiry on the basis of a division, taking as agreed the observation that both the stars and the whole heaven change position. And when he says that the whole changes position he does not mean that it changes from place to place as a whole but that the body of the heaven moves as a whole when it is spoken of in terms of the stars;[402] but it moves in its parts, as is clear from what is apprehended with instruments, namely that parts of the circle of the equator and of the circle of the zodiac, and the <stars> in the heavenly body itself rise, reach mid-heaven, and set at different times.[403] So both the fixed heaven and the fixed stars are observed to change place, and the present discussion concerns them.

'It is necessary that the change occur either with both being at rest or both moving or one at rest, the other moving', and there is no <alternative> besides these. Even though he thinks that preserving their apparent change of place is inconsistent with both being at rest, he considers as a hypothesis that both are at rest, because there have been some people, including Heraclides of Pontus and Aristarchus,[404] who thought the phenomena were preserved if the heaven and the stars were at rest but the earth moved around the poles of the equator from the west, each day making very nearly one rotation. They added 'very nearly' because of the sun's forward motion of one degree.[405] Since, if the earth didn't move (as he will demonstrate shortly,[406] but now takes as a hypothesis), it is impossible for the phenomena to be preserved if both the heaven and the stars are at rest – for how would the change of place be preserved if all things were taken to be motionless? – he dissolves this part of the division by hypothesising that the earth is motionless.

(289b7) He proves in the following way that the heaven and the stars in it do not both move. Since each star is observed to be moving in a circle, those towards the poles in smaller circles, those towards the equator in greater ones, if both the stars and the heaven were hypothesised to move, it would be necessary that both the star and the circle on which the star moves move at the same speed. And he adds the reason for the necessity: that it is observed that the stars are restored to the same place at the same time as their circles; for the fixed star which today rises together with the first degree of Aries also rises with it on the next day. And so one should imagine the motion of both as if a long-legged person seated on a horse stood up and moved along with the horse at the same speed, and the same distance was completed by both in the same time; for in this way too, with both moving and the star always occupying the same place on the circle, the star would move at the same speed as the circle. (For if the star touched the circle and moved on it with its own motion and they moved at the same speed, but the motion of the star were added <to that of the circle>, it would be necessary that the star be restored <to the same place> twice during the time in which the circle is restored once, so that the star would be observed above the earth.[407] But if the star did not touch the circle and moved faster than it, it would overtake some of it, and, if it moved slower, it would fall behind some.)

(289b13) He says that it is not reasonable that the stars move at the same speed as their circles. For greater and smaller circles in a sphere which are about the same central point or the same pole are restored simultaneously <to the same place>, but greater circles move more quickly, smaller ones more slowly. So, if the stars are restored simultaneously with their circles on which they move, the speeds of the stars will be proportional to the sizes of their circles in such a way that a star on the equator will be as much faster than a star on the always visible circle as the equator is greater than the always visible circle, since the equator is that much faster than the circle of the always visible.

But it is not just reasonable that the circles have their speeds in proportion to their sizes, it is also necessary: it is reasonable that the speeds harmonise with the sizes in ratio and proportion, since greater circles move faster; but it is necessary because it is not the case that the circles themselves are severed from the sphere and move on their
446,1 own, since that body is continuous. But since the sphere moves from the same place and is restored to the same place, all the circles, greater and smaller, on which the stars move are restored simultaneously <to the same place>. How, then, if the sphere is continuous and revolves in a circle, can one part of it already have been restored when another part is not yet restored? But if the stars moved on their own, it would no longer be reasonable that there be the same proportion between one of their speeds and another as the sizes of the circles have to each other. For it is necessary that either the circles be responsible for this sort of proportion in which the <star> moving on

the greater circle necessarily moves faster, or that one star is caused to be faster and another slower by their own selves. But if the circles are responsible, it is necessary that, if the stars were conceived as displaced, what was previously the fastest because it moved on the greatest and fastest circle become slowest by lying on the smallest and slowest circle and vice versa; and in this way the stars would no longer be moving with their own motion, but they would be carried around by the circles. But if the stars move by their own selves, some being faster, some slower, if they were divided up appropriately among the circles, the faster ones being assigned to the greater and faster circles, the slower to the slower and smaller ones, this would come about either because of some natural and primary law (*logos*) or it would just fall out this way spontaneously. But that it would come about because of some natural law is manifestly impossible, which is perhaps the reason he leaves out this part of the division. For if in the case of everything natural the greater body moves faster in its motion, as he will say shortly[408] – since a larger clod of earth moves downward and a greater fire upward faster – greater stars, when they became slower, would not harmonise with smaller and slower circles because of a natural law. But neither would it just fall out spontaneously that in all cases a circle would be greater and the motion of the star in it faster because spontaneous things are infrequent. If there were one or two cases of this kind, it would not be thought at all absurd to say that they happened spontaneously, but if, when all cases are the same, someone says that this happened spontaneously, it resembles a work of fiction (a work of fiction is something said in a way which accords with our imagination but is not in harmony with reality). So the syllogism is this:

> (i) What happens spontaneously and by chance is not everywhere and in all cases;
> (ii) therefore, the proportion between the stars and their circles is not spontaneous or by chance.

And he has set out another syllogism of the following kind:

> (i) What is everywhere and in all cases is natural; 447,1
> (ii) what is natural is not by chance or a matter of chance;
> (iii) so, if the proportion in the stars to their circles is everywhere and in all cases, it will not be a matter of chance or by chance.

Alexander interprets the words 'and[409] at the same time there is no chance in natural things, and what holds everywhere and in all cases is not a matter of chance'[410] as follows:

> What is by chance or a matter of chance is not found in natural

> things, since what happens naturally happens usually, not rarely. Nor is what is by chance or a matter of chance found in necessary things, since what is everywhere and belongs to everything is necessary, since the opposite of the necessary is impossible. But things in the heaven are put together naturally and by necessity, whereas things which are by chance or a matter of chance are infrequent and rare; consequently, it is not reasonable that things just fell out this way by chance.[411] So if the stars are restored simultaneously with their circles (as is clear from the phenomena), it follows for those who say that the stars move <on their own> that they move at the same speed as the circles on which they move. And it follows from this either that the stars are moved by the necessity of the circles and not on their own or that, if they move on their own, then the proportion of the stars to their circles is preserved either naturally or spontaneously and as a matter of chance. So if both of these have been proved impossible and these follow from the <hypothesis> that both the circles and the stars move, it is clear that the hypothesis is impossible.

But, because Aristotle's words can also apply to the spheres, Alexander says that he[412] has called the spheres circles.[413] For the stars in the planetary spheres are observed to be restored simultaneously with the revolution of the spheres with the result that each of them would move at the same speed as the sphere on which it is. And some of the spheres are greater, some smaller, since the outer and containing spheres are greater than the inner and contained ones. And the words 'there is nothing absurd in the circles having their speeds proportional to their sizes; rather it is even[414] necessary' would be equivalent to 'it is necessary that the spheres have their speeds proportional to their sizes'; for the greater sphere moves in the greater time; accordingly the sphere of Saturn <makes a complete west-east revolution> in thirty years, that of Jupiter in twelve, and other spheres in other times.[415]

But perhaps Aristotle says that the circles have their speeds proportional to their sizes because the greater circles move faster, since they are restored simultaneously with the smaller circles. But Alexander says the contrary in the case of the spheres, namely that the greater sphere makes its motion in a greater time. <Aristotle> is
448,1 speaking generally about the circles which are restored simultaneously, but the planetary spheres are not restored simultaneously, as Alexander himself recognises as he goes on.[416]

Alexander also adds this:

> If both the circles and the stars move and they are restored at

> the same time, one of the motions is obviously pointless, but there is nothing pointless in what happens naturally.

And I wonder how he can say this, since even if the circles and the stars are restored simultaneously, they have their own substances and powers, and they are active in accordance with them.[417]

> **289b27-30** And again, if the circles remained fixed [and the stars moved, there would in the same way be the same unreasonable consequences, since it would follow that outer[418] things would move faster] and their speeds would accord with the sizes of their circles.[419]

He has proved that neither can both the circles and the stars be at rest nor can they both be in motion. Now he proves that it is also not reasonable that the stars should move and their circles remain fixed, since the same unreasonable consequences follow in the same way. For if the stars are assumed to be moving and their circles at rest, it will result also in this circumstance that the star on the greater circle moves faster and that the speeds of the stars stand in proportion to the sizes of the circles on which they move. But this is impossible, since it is not possible for this to happen naturally, since greater magnitudes naturally move in similar ways faster than smaller ones.[420] So it is necessary either that the circles are responsible for this kind of proportionality (and so the stars will not be moving on their own since a smaller star would not be moving on a greater circle[421]) or that it just fell out this way spontaneously and as a matter of chance, which is an unreasonable thing to say in the case of natural things and most of all in the case of divine ones. So there is no way in which it is possible that, with <only> the stars moving, the speeds of the stars could stand in proportion to the sizes of their circles. Consequently 'there would in the same way be the same unreasonable consequences' on this hypothesis too.

When Aristotle says that it would follow that outer things (that is, those which are further from the poles <of the cosmos>) would move faster and their speeds would accord with the sizes of their circles, Alexander makes the following remark:

> He could not be referring to the simultaneous restoration of the planets since these are not restored simultaneously, but each is restored <to the same place> in different times according to its own motion. But if these are not restored simultaneously, their different speeds will not accord with the difference in the magnitude <of the circles>[422] on which they move. Consequently, it 449,1
> is reasonable, as we have said, that he gives these proofs in the present case for the stars in the fixed sphere.

289b30-290a7 So since[423] it is not reasonable either that both move [or that just one does, it remains that the circles move and the stars are at rest and, being fastened in the circles, are carried around. For only in this way does nothing unreasonable result. For <first> it is reasonable that in the case of circles fastened about the same central point the speed of a greater circle will be faster, since, just as in other cases, the greater body moves faster in its own motion, so too in the case of circular bodies; for of arcs cut off by things from the central point[424] that of the greater circle is greater, so it is reasonable that a greater circle <and a smaller> will be carried around in an equal time. <Second> it will result both from this and from the fact] that the universe has been proved to be continuous[425] that the heaven is not torn apart.

There are four parts of the division which states in how many ways it is possible to preserve the apparent change of place of the stars. And he has proved that it is not reasonable either that both the circles and the stars be at rest or that both move or that just one of them, obviously the stars, moves (he has just proved this about them). He now reasonably adds that 'it remains that the circles move and the stars are at rest and, being fastened ..., are carried around'. For this is the remaining part of the division. And, he says that only in this way 'does nothing unreasonable result'. And he next proves that the consequence of the other hypotheses – its being unreasonable that the speeds of the stars are proportional to the sizes of their circles – does not follow from this hypothesis, but what follows from this hypothesis is not only that it is not unreasonable that, in the case of circles fastened about the same central point, the speed of a greater circle will be greater, but also that it is necessary, and this in two ways. First (i) because just as in the case of other natural magnitudes the greater body moves with its own natural motion faster than a lesser body does with its, a greater clod of earth faster than a smaller one, and a larger fire than a smaller one, so too in the case of circular things, the greater moves faster. For if straight lines are extended from the central point of homocentric circles to the circumferences of the circles, they cut off similar arcs, e.g. a fourth or a semicircle or whatever; and when the circles move, if one of the straight lines from the central point is also conceived to move along with the circles, it will coincide with the other straight line conceived as remaining fixed, and then in the same time each of the similar arcs will have moved in itself, one a greater distance, the other a less.[426] Consequently it is reasonable that the whole circles, greater and less, will be carried around in an equal time. So it follows necessarily from the speeds being proportional to the sizes that greater and smaller circles are restored simultaneously <to the same place> in an equal time and further (ii) that the heaven is not torn apart. This is true of the heaven principally because the universe has been proved to be continuous, and it also results because of the proportionality of the motions of its parts to their sizes. For if a greater circle, e.g. the equator, moved at the same speed as a smaller one, e.g. the always visible circle, since 450,1 they would not be restored simultaneously in the same time, the continuity of the universe would be torn apart.

One should note that the *te* in *to te mê diaspasthai ton ouranon* <that the heaven is not torn apart> is answered by the *te* in *to te gar thatton einai tou meizonos kuklou to takhos* <the speed of a larger circle will be faster>.[427]

Alexander says the following about his words 'fastened about the same central point':

> It is possible to understand this as referring to the spheres <of the planets>, and what would be eliminated by this is the view of those who explain the motion of the planets in terms of eccentricity and try in this way to preserve the phenomena.
>
> And what comes next, 'for of arcs cut off by things from the central point that of the greater circle is greater', would be spoken about straight lines <from the central point>. However, although it is true that the arcs of outer circles are cut off greater, this is not evidence that it is reasonable that a greater circle moves faster. Rather, if it does move faster, it does so because of the fact that in the case of natural motion a greater body moves faster.[428]

But <Alexander> also remarks that these words are not suitable to the planetary circles but to the circles in the fixed sphere:

> For it is true in the case of circles in the fixed sphere that a greater circle is restored <to the same place> in an equal time as a smaller one is, but it is not true in the case of the planetary spheres, either with respect to their own proper motion[429] (since they are restored <to the same place> in a different time) or with respect to their motion together with the universe (since each is left behind by that motion to the extent that in its own motion it moves in the direction which is reverse to that of the universe; consequently the stars in the planetary spheres are observed not to be restored <to the same place> simultaneously with the fixed stars with which they happen to be in conjunction, but they are left behind).

<Alexander> understands the words 'fastened about the same central point' to apply to a central point in the strict sense; he says they are equivalent to 'which are in the same sphere'.[430] But perhaps, since

both straight lines extended from the central point and arcs extended from the pole cut the greater and smaller circles proportionally, although he now needs the arcs extended from the pole, he indicates them by referring to things from the central point. For the theorem is clearer in their case and is presented in plane geometry prior to the study of spherics.

But why does Alexander say that this proportional sectioning is not evidence that it is reasonable that a greater circle moves faster? For if the arc of the greater circle and its analogue in the smaller one move through the distances between the straight lines from the central point, one of which is greater, the other smaller, in the same
451,1 time, why isn't that clear evidence? And I just do not know why Alexander does not think that the things from the central point are meant to be straight lines rather than planes drawn through the central point to divide the circles proportionally, even though Aristotle says 'from the central point' and not 'through the central point' and what are divided proportionally are circles, not spheres.

When Aristotle says 'it results both from this and from the fact that the universe has been proved to be continuous[431] that the heaven is not torn apart', Alexander correctly recognises that this text also makes clear that he is speaking about the sphere of the fixed stars. For <Aristotle> speaks of something continuous, and the planetary spheres are not continuous; rather they touch one another, and there could not be a tearing apart in the case of these things which move in different ways.[432] These things are well said by Alexander, but we should consider what he says next:

> Since the body of each sphere is continuous, there would be a tearing apart in the case of a sphere if the stars were moving[433] while the spheres remained fixed; for then their bodies would no longer be continuous, if the stars, which are parts of them, changed from place to place.

But perhaps this is not what is being set forth now, since it has already been proved[434] that the stars do not move while the spheres remain fixed. Rather what is being set forth is that the proportionality of the speeds of the circles in relation to their sizes is the cause of the fact that, since the greater and smaller circles are simultaneously restored <to the same place> in the same time, the heaven is not torn apart. Accordingly here there has nowhere been a reference to a star. Alexander persists in this same opinion when he says:

> It is possible to understand this as applying to all the spheres, not because they are continuous with one another, but because they are continuous with the stars in them, and each of the spheres would be torn apart if the stars in them moved.

However, he inferred correctly a little while ago that it is impossible for the planets to be restored <to the same place> simultaneously with each other or with the fixed sphere.[435]

So now Aristotle has concluded that, since the other parts of the division are unreasonable, there remains the one which says that 'the circles move and the stars are at rest'. But it is worth inquiring what these circles are. For the equator and the zodiac, since they bisect the sphere, are reasonably thought to be inherent and to determine their hemispheres in a substantial way, but perhaps the other circles exist <only> in our thought and we draw circles for each of the stars with the axis as central point and a radius. Or perhaps the circles of the motions do not exist only in our thought, but, since the whole sphere always moves around the same poles (those of the equator), each of the stars in the sphere, always remaining in the same position, being carried around by the whole sphere, and always being at an equal distance from the axis, has a circular motion. And so, even if these 452,1
circles have not been engraved in the sphere,[436] since every motion covers some distance, and <in this case> the motion is circular, it is also necessary that the distance in which the stars which are equally far from the axis have been fixed by the demiurge-god be circular in a natural way. And these are the things which Aristotle says have their speeds proportional to their sizes.[437]

290a7-24 Furthermore, since the stars are spherical, [just as others also say in agreement with us, since they generate the stars from that body,[438] and there are two ways in which a spherical body moves on its own, rolling and rotation,[439] if the stars move on their own, they should move with one of these two motions. But they are not observed to move with either.

290a12 For, if they rotated, they would remain in the same place and not change place, which they are observed to do and everyone says they do.

290a13 Furthermore it is reasonable that all the stars move with the same motion, but of them only the sun seems to rotate when it rises and sets,[440] and it doesn't do so on its own, but it seems to because of its distance from our vision. For when our vision is stretched out a long way it rolls around because of its weakness.

290a18 This is perhaps also the reason that the fastened[441] stars appear to twinkle, but the planets do not twinkle; for the planets are close so that vision[442] reaches them while it is strong, but when it is directed at the stars which remain[443] it trembles because of the length, because it is stretched out too far. And its shaking makes the motion seem to be the star's;] for there is no <visible> difference between vision and the thing seen causing the motion.

After his demonstration by division concerning the motion or absence of motion of the stars he adds another kind of demonstration that the stars do not move. He hypothesises that they are spherical, and now he makes the hypothesis plausible both on the basis of the opinion of others who think the same way and on the basis of the necessity that the stars have the same shape as the heavenly body since they have the same substance. But in a little while he will put forward <the proposition> that the shape of the stars is spherical on its own and prove it.[444]

So, he hypothesises that the stars are spherical and assumes in advance that there are two motions of spherical bodies insofar as they are spherical, rotation and rolling. Rotation occurs when a sphere turns around its own axis in the same place; rolling when it <turns around its axis and> changes place. And, having proved next that neither of these motions is suitable for the stars if the spheres are at rest,[445] he has shown, as Alexander says,[446] that the stars themselves do not move.

(290a12) And he proves that they do not rotate from the fact that they are observed to change position and not remain in the same place. For, if what rotates remains in the same place (that's what rotation involves), but the stars do not remain in the same place but change place (as the phenomena and the common opinion of all people bear witness), it is clear that it follows in the second figure that the stars do not rotate or – more accurately – that the stars do not have that change of place which we see in them because they are rotating.

(290a13) Next he dissolves the objection which says that the sun is observed to rotate, so that it is reasonable that the stars also move in this way. And he dissolves this objection in several ways: first from the fact that it is reasonable that all things which have the same substance move with the same motion, but only the sun is seen
453,1 rotating; second from the fact that the sun is not observed rotating everywhere but only when it rises and sets; and third by explaining that the cause of the imagined rotational motion of the sun lies with us and not with the sun. For, he says, when vision is stretched out a long way, it becomes weak and trembles because of the length, since weak things are affected by trembling when they move. This trembling and shaking and rolling around of the viewer is thought to be a rotating of the thing seen, since, as far as an appearance of this kind is concerned, there is no difference between hypothesising that vision moves or that the thing seen does. (For those who sail along[447] coasts believe that the earth and the mountains are moving although it is their vision which is moving, and it transfers its own affection to what it sees.)

(290a18) He also confirms that this is the reason why the sun appears to rotate on the basis of the fact that the fixed stars appear to twinkle. He calls these stars 'fastened' either to contrast them with the stars which are thought to wander or because they are fastened in the fixed sphere, which is the topic of discussion. The planets are not thought to be subject to twinkling. For the planets, he says, are close so that vision reaches them while it still has control over itself and is powerful, so that vision is not affected by anything and predicates no affection of the planets. But when it is stretched out towards the fixed stars, which are further away and which he calls 'the stars which remain' because they are fixed, it is stretched out and trembles, being weakened by the great length, and its shaking makes there appear to be rotation in the case of the sun and twinkling in the case of the fixed stars.

But it is manifest that twinkling is also observed in the case of the sun. Why then, one might ask, is rotation only observed in the case of the sun and not also in that of <stars> beyond the sun, in being stretched out towards which, since they are further, vision would tremble and shake more because of greater weakness? In response to this difficulty Alexander says:

> Perhaps this happens partly because of the brightness of the sun, and perhaps partly because of the earth so that the sun is seen as rotating when it is alongside of[448] the earth (since it is seen rotating both when it rises and when it sets). For it somehow turns out that its two motions, its revolution with the universe and its own reverse motion, are seen by being viewed alongside of the stationary earth so that it seems to tremble; and therefore we see it trembling only when it is rising or setting. And this would happen only in the case of the sun, because it is the largest and brightest and most evident of the nearby <stars>.
>
> When <Aristotle> said 'for when our vision[449] is stretched out a long way' he is not referring to an emission of rays when we see, since he does not believe in them. Rather he has now called 'vision' the transparent medium, which, because it communicates a form to us through a great intervening distance, does not receive the form of the thing seen in the same way as when <the thing seen> is near to our vision, nor does it master it because it is transmitted from a long way.
>
> And Aristotle also means to apply[450] the words 'so that our[451] vision reaches them while it is strong' to the transparent medium.

But Alexander perceives that it is exceedingly implausible that <Aristotle> would call the transparent <medium> vision and say that it rolls around because of weakness. He says:

> Even if he were speaking about vision and rays, he would not be expressing his own view, but rather <he is saying this> because

> the same affection is produced whether the transparent medium is what is affected by the object perceived and communicates <it> to vision or whether rays are sent out from vision. But the latter view is more common because the mathematicians talk that way; for in many places he uses the view, even though it is not his own opinion, since he will prove in *De Sensu*[452] that things cannot be this way.

When he says 'the planets do not twinkle' despite the fact that the sun is one of the planets and does twinkle,[453] he might mean the planets other than the sun or he might be calling only the five planets besides the sun and moon planets. And it is clear that he says this with respect to what is true in most cases. For the name attached to Mercury[454] makes clear that among the planets it twinkles, and it is thought by some people that Venus twinkles as well. And it is better to refer these things more to the special characteristics of the stars, such as their colours.[455]

Those who strive contentiously to put Aristotle in a different camp from Plato and say that this doctrine of theirs is also a matter of disagreement allege that Plato says the fixed stars have two motions, one in which they are carried around by the heaven and one which is their proper motion of spinning and rotating about their own central points, whereas Aristotle eliminates rotation. So, if you agree, let us look at the opinion of each, and first let us hear what Plato says. He writes the following in the *Timaeus*:

> <The demiurge> fashioned[456] the greatest *idea* of the divine <kind of living thing> from fire so that it would be the most beautiful and the brightest to see. Likening it to the universe, he made it well rounded and set it into the wisdom of the supreme which accompanies it. He distributed it in a circle around the whole heaven to make the heaven a genuine cosmos embroidered by it[457] in its entirety. And he attached to each thing of the kind two motions, one uniform rotation in the same place, each always thinking the same things about the same things,[458] the other a forward motion, each being dominated by the revolution of the same and similar. But he made each unmoving and stationary[459] in terms of the <other> five motions[460] in order that each of them would also[461] be as good as possible. For this reason all the fixed stars, which are divine[462] and therefore turn in the same place,[463] came to be and remain forever; and the stars which turn and wander in this way came to be in the way which was described previously.[464]

Here Plato says that each of the fixed stars has two motions, one its own motion about its own central point while remaining in its own place, the other a forward motion brought about by the heaven which contains them.

It is clear that Aristotle eliminates rotation; however, he does not eliminate it as not existing at all among the <stars>, but as not existing by itself with the spheres being fixed, and also as not being the cause of the apparent change in position <of the stars>. For he says, 'if[465] they rotated, they would remain in the same place and not change place,[466] which they are observed to do and everyone says they do'. And in the sequel[467] he says that nature, as if purposively, eliminated all of the things by means of which they could progress on their own and left them sphericity, which he agrees is more serviceable for internal motion. He says, 'for in this way they can move most quickly and most perfectly occupy[468] the same place', and both of these features are appropriate to things that rotate. But at the end of the discussion he says, 'so, since it is necessary that the heaven have an internal motion, but that the other stars[469] not progress on their own, it is reasonable that each be spherical, since most of all in this way, the one will move and the other will be at rest'. Thus he eliminates not progressing on one's own, but not motion in general. And he says that the heaven moves since the change of place of its parts is perceptible because of its magnitude, but that the stars are at rest since the change of place of their parts which involves motion about their own central points is not seen and their entirety always occupies the same position in the heaven.[470]

And in general, if the stars are natural things and of the same substance, the fifth, as the heaven, and they have been given the same spherical shape, how can they not have the same motion, at least if they are natural things? And if, as he says in the sequel,[471] heavenly things, which obviously have soul, share as wholes in action and life, how can the stars not move on their own, since, even if they are parts of heaven, they are parts which have their own wholeness and determinate individuality. Furthermore, if a spherical shape is suitable to rest, no body would naturally move in a circle.

We should also attend to the whole argument. He says that since the stars are spherical, if they moved on their own, the motion which they would be observed to have would be either rotating or rolling. But their motion is not rotation, since what rotates remains in the same place. Nor is it rolling, as he will prove.[472] Therefore, they do not have this motion which involves change of place on their own. However, he has not proved that the stars have only this motion of changing place which is caused by the heaven and have no other, unless he is thought to eliminate rotation completely by proving that the sun does not rotate when it is near the horizon. But if he eliminated the rotation in itself of the stars, he would not have eliminated rotation from their progression by saying, 'for if[473] they rotated, they would remain in the same place'.[474] But since the sun is thought to rise and set while rotating because our vision rolls around in such a way that it seems to change place while rotating, he has dissolved the objection based on the sun on this ground, by supporting the view that what rotates remains in the same place and proving that the apparent rotation of the sun in its progression in rising and setting is not to be attached to the sun itself but to our vision.[475] But one should also hear the best of the astronomers, Ptolemy, who says in the second book of the *Hypotheses*:

> So it is more reasonable that each of the stars causes motion in its own place, because this is both their power and activity, but about its centre and again uniformly and in a circle; for it is right that what achieves this <rotation> in the surrounding structures belong to the star first.[476]

> **290a24-9** However, it is evident that the stars do not roll either. [For it is necessary that what rolls turn, but the so-called face of the moon is always visible. So, since it is reasonable that if they moved on their own, they would have their proper motion, but they are not observed to move in this way,] it is clear that they do not move[477] on their own.

He has proved that the apparent motion of the stars is not rotation – 'for if they rotated, they would remain in the same place' – and he has
457,1 dissolved the objection based on the sun. Next he proves that the stars do not make their apparent change of place by rolling. For even if they do change place in the way rolling things do and even if rolling is proper to a spherical body, he proves that they do not change place by rolling from the fact that the apparent face of the moon is always seen to have the same position; but this would not happen if the <body> in which the shape is present revolved and rolled over and was different at different times. It is clear that this argument proceeds on the basis of an axiom which says that what holds of one thing also holds of all; and the axiom is reasonable in the case of divine things. And if the visible face is a certain substantial difference of the lunar body and either contains the multiple differentiation of coming to be in a paradigmatic way, as Iamblichus says, or, as others suppose, it has one part well lighted, the other darker in substance because of being the middle term between heavenly and sublunary things, what <Aristotle> says seems strong as demonstration. For how, if the whole body rolled, could the face, which is continuous with the whole, retain the same position? But if, as some people say,[478] it is a kind of mirror image of the earth or the sea or the mountains, it is possible, Alexander says, that, although it rolled, the image would remain the same because the things which are reflected remain the same and the moon has the same suitability for receiving such an image in all its parts,

both because of the nature of its body and because of the uniformity of its shape. However, if a reflected image comes to the thing reflected because of the bending back of rays by the mirror, as many people think, the same thing would not be reflected back to the viewer from every place in the way that now the face of the moon appears the same to a person seeing it from the far south or from the far north or from the east or the west. However, if from different places different effluences are always ascending from sublunary things and taking on the kind of structure in the brightness of the moon which some people say the reflection of images takes on in mirrors, it would be possible for the same image to be seen from every place, even though the moon rolled and rotated.

One should notice why Aristotle did not also use the argument from the face of the moon in the case of rotation: for since what rotates has a different position at different times, <the moon> would not present the same image of its face if this differentiation of its body were in the moon in a substantial way. Or is it possible that the 458,1
sphere in which the moon is fastened moves in the opposite direction of the moon's rotation and carries it around in such a way that there is produced a uniformity in the position of the face?[479]

He has proved that it is not possible for the stars to have the motion which they are observed to have if they either rotate or roll; and, since it would be impossible for them to move on their own in any other way, he finally concludes that 'they do not move on their own'.[480]

290a29-b11 In addition to these things it would be unreasonable [that nature did not provide the <stars> with any organ for motion (since nature does nothing by chance) and that nature should provide for animals but overlook such honourable beings. But it seems that nature, as if purposively, eliminated anything by means of which they could progress on their own and set them at the greatest distance from things which have organs for moving.

Therefore it would seem reasonable that the whole heaven and each of the stars be spherical. For the sphere is the most serviceable shape for internal motion (for in this way they can move most quickly and most perfectly occupy the same place), but it is the least serviceable for forward motion. For it is the least similar to things which move[481] on their own since none of it is detached or projecting, as in a rectilinear figure, but it stands at the greatest distance from the shape of bodies which move forward.

290b8 So, since it is necessary that the heaven have an internal motion, but that the other stars[482] not progress on their own, it is reasonable that each be spherical,] since most of all in this way, the one will move and the other will be at rest.[483]

He has proved that the stars do not move on their own in the way in which they are observed to move because they do not move in that way because of either rotation or rolling. He now proves again on the basis of another idea that they do not change place on their own. The syllogism amounts to this:

(i) Things which change place on their own have detached and projecting parts in the way that animals have feet;
(ii) the heaven and the stars are spherical and have no detached or projecting parts;
(iii) and the conclusion in the second figure is that the heaven and the stars do not change place on their own. (But it is clear that the heaven does not change from place to place at all, while the stars do change place but not on their own.)

Since he has based his argument on a similarity to the other animals, he first proves that it is unreasonable, since nature does nothing which is pointless, that it would furnish less honourable animals (which he has specifically called animals because they are familiar <cases of animals>)[484] with organs for changing place but neglect the more honourable ones. If this is absurd, it will rather be true and fitting that nature, as if purposively, eliminated from the heaven and the stars anything by means of which they could progress (that is, change place) on their own, and that it set them at the greatest distance from things which have organs for changing place and made the heaven and each of the stars spherical because, he says, spherical shape is most suitable for internal motion. For in this way the heaven would move both most quickly and uniformly, and be a measure of all motion, and it would most perfectly occupy the same place since there would be nothing outside of it. It moves most quickly because, of lines containing an equal area, the round one is least,[485] but of things which are moved by the same power, that which traverses the lesser perimeter moves <through it> faster.[486] Moreover, because its surface is smooth, similar at every point, evenly balanced, and, as Plato says, 'it always stands on the smallest base',[487] what is spherical moves most easily and has a uniform motion because it does not offer resistance at angles, and it is not hindered, and it does not move non-uniformly. For figures with angles have a non-uniform motion because of the extrusion of the angles and the non-uniformity of their shape, and they revolve faster when they are resting on an angle (just as what is spherical always moves around one point) but they move most slowly when they rest on a plane surface. But also a rotating spherical figure always occupies the same place because it has no extrusion which is forced to take up and leave different places at different times in the revolution. So in this way spherical shape is most serviceable both for internal motion (which occurs when parts take each other's place) and for resting in the same place, but it is most unserviceable for progression forward, since it is entirely dissimilar to the things which change place on their own. It is dissimilar because things which are rectilinear and not spherical have certain projecting and detached parts which they use as organs for changing place, with the result that they do not touch what lies under them with their most authoritative parts,[488] but if a sphere were to change place and roll it would use its whole body for the change of place. So if spherical shape is least serviceable for progression forward, it is clear that rolling does not attach to it naturally.

Alexander raises the question how <Aristotle> can say that the fixed heaven occupies the same place when it is not *per se*[489] in place at all?[490] And he says that <Aristotle> is now calling extension place in a more general sense. But what kind of extension? If he means the extension of the body itself, this thing which exists in the extension of its own particular body also belongs to things which change place; but if we are calling space extension, we are now agreeing to what we try to eliminate in our discussions of place.[491]

(290b8) Finishing up his discussion, Aristotle says, 'so, since it is 460,1
necessary that the heaven have an internal motion' (he either means motion in the same place – because he agreed that the heaven <always> occupies the same place – or, in case it is not in place, he is making a contrast with motion in place), 'but that the other heavenly bodies (that is the stars)[492] not progress on their own' towards the front in the way in which things which move forward and change place do, 'it is reasonable that each be spherical', since most of all in this way the heavenly body will have an internal motion, which is also what we see, and the stars will always occupy the same position.

There are also texts which say 'the one will cause motion and the other will be at rest'.[493] And this text is clearer since the heaven causes the stars to revolve and the stars remain in the same place.

And notice that he does not say simply that it is necessary that the stars do not move at all but that 'they do not progress on their own' – which is also true of the heaven.[494] For he says that nature purposively eliminated from the stars the organs by means of which they could progress on their own. Since both the heaven and the stars remain in the same place and both move in a circle, and we see both the circular motion of the heaven because of the revolution of the stars and the remaining of the stars in the same place in the heaven, he characterises both on the basis of what we see when he says, 'the one will move and the other will be at rest'.[495]

And one should note that this whole argument is aimed at proving that this apparent change of place which the stars are observed to have each day when they rise, reach mid-heaven, and set is not a matter of their progressing on their own but of their being carried around by the heaven while always resting in the same place in the heaven. But since the bodies of the heavenly spheres are not seen by us because they are transparent, but we only view the stars, what makes it clear that there are heavenly bodies at all and that the stars do not move in the *aithêr* with their apparent motion on their own? Why has Aristotle, who always provides demonstrations based on clear phenomena, failed to investigate this? Or has he not failed, but rather proved clearly <that the stars do not move on their own> when he proved that the speeds of the stars are proportional to the sizes of the heavenly circles in which they are carried around and said that it is necessary – as also in other cases – that the greater body move the same distance faster?[496] But if some smaller stars are observed to be faster than greater ones and to complete greater circles in an equal time, it is clear that the proportionality involves the sizes of the bodies in which they are carried around and that these are the heavenly things the speed of which is proportional to the size. In general, if it has been proved that the stars do not have their apparent motion on their own, but that they are carried around by other spherical bodies, it has also been proved by Aristotle that there are heavenly spheres. And he proved at the start of the treatise[497] on the basis of the fact that there are two simple lines that there are two simple motions and on the basis of this that there are two <types of> simple body which move with the simple motions, the body which moves in a circle and the body which progresses in a straight line. This motion of the stars when they move from place to place is not circular in the strict sense, even if it occurs on a circle. For circular motion <in the strict sense> involves remaining in the same place; therefore, it is necessary that there be a body which is more whole than the stars, which are carried around from place to place.

And so Aristotle has demonstrated in many ways the existence of the heavenly body. But it should be asked what there can be between the stars. For there isn't any void, since this has been proved not to exist.[498] So if there is body <between the stars>, either this is the eternal heavenly thing we are seeking or it is something which comes to be and is destroyed like air or some aitherial *hupekkauma*.[499] And how is it possible that these things which differ so much in substance, I mean the stars, which are eternal, and this intermediate body, if it comes to be and is destroyed, have the same place naturally? And how is it reasonable that the entireties of bodies which come to be and are destroyed are continuous and unified but the eternal body is divided and torn apart and does not have one corporeal entirety?

On the basis of what is proved in astronomy the existence of the heavenly body can be proved through the motion of the nodes involved in eclipses of the sun and moon. For these two luminosities move on circles which are oblique to one another, and the nodes involved in eclipses are the diametrically opposite intersections of the circles; for when both of these lights are at the same intersection there is an eclipse of the sun, but when they are at diametrically opposite intersections the moon is eclipsed. So, since the intersections are observed to move uniformly, with the result that eclipses do not always occur at the same position, but the sun has been shown always to move on the same circle, the one through the middle of the zodiac, then if the moon moved on its own obliquely to that circle, it would still intersect the solar circle at different points at different times, but it would no longer move in a circle but would describe a helix.[500] But this would be absurd, at least if it ought to have a simple and circular motion because it is made of the fifth substance. So the moon has an oblique circle, and the nodes change place in accordance with the motion of the circle. However, the circle does not exist or move on its own, but it has its being in a sphere and moves along with it. Therefore there is a lunar and a solar heaven,[501] and if these exist, so do the heavens which contain the other stars.

However, one could probably also demonstrate this by means of other, more persuasive demonstrations, but this much should be added to what has been said. If the so-called fixed sphere were really fixed and one did not accept the observation of Hipparchus and Ptolemy[502] concerning it, according to which it also moves in reverse one degree in a hundred years, it would have one motion and that a simple one, and the stars in it would have two motions, their own rotation and the rotation of the universe, and the planets would have three motions, their own, that of the spheres containing them, and that of the universe.

However, when I was present in Alexandria, our master Ammonius observed Arcturus using an armillary sphere[503] and found that it had moved forward so much with respect to its position at the time of Ptolemy as to require that it had one degree of contrary motion in one hundred years. So perhaps it would be truer to say that the starless sphere which contains all <the spheres>, of which it seems there was no knowledge at the time of Aristotle, carries around all the other <spheres> with its single simple motion from the east;[504] that the sphere which we call fixed has two motions, one which is from the east and is that of the universe, the other which is from the west and is its own; that the stars in this sphere have these two motions and their own rotation; and similarly that the succeeding spheres have the same two motions and the stars in them have the same three motions.[505]

It should also be mentioned that when Aristotle says that it is unreasonable that nature 'should provide for animals but overlook such honourable beings', Alexander says that he has referred to things which come to be as animals because the ability to perceive is peculiar to animals, but this faculty of soul does not exist among divine things, just as the nutritive faculty does not, with the result that they are not animals having soul except homonymously.[506] It would certainly be surprising if the divine <body> were perceptible and tangible, as he says, but could not perceive. Is it because it is better for a body not to perceive? However, not perceiving belongs to the last and least honourable bodies, those without soul. Perhaps then one should eliminate <from divine bodies> the senses which involve matter and are most passive, smell and taste, but assign them the other very precise ones; for they do not touch one another without perception, and they see all things and hear all things.[507]

[Chapter 9]

> **290b12-29** It is evident from these things that those who say that a harmony is produced when they move [because the sounds are harmonious speak in an ingenious and clever way,[508] but what they say contains no truth. For some people think it is necessary that a sound be produced when such large bodies move, since things in our world, which do not have bulk equal <to theirs> and do not move with such speed, <produce sound>. But when the sun and moon and, in addition, stars of such a number and size move with such speed, it is impossible that some sound of incredible magnitude not be produced. Hypothesising these things and that the speeds (<inferred> from the distances) have the ratios of harmonies, they say that, when the stars move in a circle, the sound is harmonised. But since it seemed[509] unreasonable that we do not hear this sound, they say that the reason for this is that the sound already exists at birth so that there is no way to distinguish it from an opposing silence. For sound and silence are distinguished relative to one another; consequently] the same thing happens to human beings as happens to blacksmiths, who think there is no difference because of their habituation <to the sound>.

The project now is to refute those who say that a sound, a harmonised one, is produced by the motion of the stars. The issue is in itself appropriate to discussion of the motion of the stars, and it has as a consequence the destruction of the view, refuted in the preceding, that the stars move[510] on their own. For if it were the case that if they moved on their own it would be necessary that they make a sound, but it were proved that they do not make a sound, it is clear that they wouldn't move in this way either. And this is the view of the Pythagoreans; they said that Pythagoras perceived the harmonised sound produced by the motion of the heavenly bodies.

Aristotle first describes the view in a plausible way and then tries

to refute it.[511] He says that these people hypothesise that a sound is produced by the motion of the heavenly bodies and they make their hypothesis plausible from the fact that when bodies in our world move a sound is produced, although bodies in our world do not have a size equal to theirs and they do not move with such speed. By much 464,1
more then, when the sun and moon and the other stars move in such multitude and size and speed that they traverse such a great heavenly distance in twenty-four hours, 'it is impossible that some sound of incredible magnitude not be produced'.

So, they hypothesised plausibly in this way that sound is produced, they apprehended that the distances <of the spheres> from the earth are in harmonious ratios, and they inferred from the distances that their sizes are in the same ratio as the distances, since things that contain are always greater than the things they contain and they are as much greater as the distance of the container from the contained. And then they discovered that the speeds of the motions are in the same ratio as the sizes because a greater body moves the same distance faster, as <Aristotle> himself says,[512] and as much faster as it is greater. And since faster motions make higher pitched sounds, slower ones lower pitched sounds, but the ratios of harmonised things exist in high and low pitch, they, having found these ratios in the distances – and because of the distances the same proportionality is preserved in the sizes, which are proportional to the distances and because of the sizes, the same proportionality is preserved in their speeds[513] – reasonably 'say that, when the stars move in a circle, the sound' …[514] 'is harmonised'.

And notice that the distances and sizes of the sun and moon and their ratios are apprehended through geometric demonstrations, as we learn in the astronomical theorems about sizes and distances. And if the ratios of the distances and sizes of the other stars higher than the sun have in fact been apprehended by them (although up until the present other people do not know them), then, since the sizes follow the distances in proportion, and the speed of the motions follows the sizes in proportion, and the high or low pitch of the sounds follows the motions, it is necessary that the sound emitted be completely harmonised.

But since it seemed unreasonable that we do not hear the sound, they dissolve this objection by <saying> that the reason for this is that the sound already exists when we are born, and it is continuous and without remission. We do not perceive the sound because of habituation and familiarity and because it is continuous, since sound and silence are distinguished best by juxtaposition with one another, as are motion and rest, white and black, and the other opposites. Thus we do not perceive the sound because there is no way to distinguish it by comparison with an opposing silence. Consequently, just as the banging does not disturb blacksmiths because of habituation, nor do 465,1 they think there is any difference when they are hearing the banging, so too and still more, that sound is not perceptible to us because of habituation.

These, then, are the things which, because they are plausible, he says are spoken in an 'ingenious' – that is, attractive – and 'clever' – that is, inventive and thoughtful – way.

290b30-291a28 As we said before, these things [are indeed spoken tunefully and musically, but it is impossible that things are this way. For it is not just anomalous that we hear nothing (and they do try to resolve[515] the reason related to this), it is also strange that nothing is affected apart from perception. For excessive sounds even shatter the bulks of bodies without soul, for example, the <sound> of thunder splits stones and the strongest bodies. With so many things moving and the sound passing through being relative[516] to the moving magnitude, a magnitude which is many times as great must reach here and <its> strength must be incredible in its force. It is reasonable <to conclude> that we do not hear <anything> and that bodies are not observed to be affected forcefully because there is no sound.

291a6 At the same time the reason for these things is clear and it gives witness that the things we have said are true; for what was a difficulty and made the Pythagoreans say that there is a harmony in the moving things is evidence for us. For things which themselves move make a sound and an impact,[517] but whatever is fastened or inherent in a moving thing, such as the parts in a ship, cannot make a sound, nor can the ship itself if it is moving in a river. However, it would be possible to say the very same things: that it is strange if, in moving, the mast or the stern of so large a boat, or again the ship itself, when it moves, does not make a sound. But if something moves in something which is not moving it makes a sound, but it is impossible that it make a sound if it is in something moving with which it is continuous and it does not make an impact. And so here one should say that if the bodies of these things moved in a mass of air diffused through the universe or a mass of fire – as everyone says they do – it would be necessary that they make a sound of enormous size, but if this occurred, it would <be necessary that it> reach here and shatter things. So, since this is not observed to happen, none of these things has either a motion involving soul or a forced motion. It is as if nature foresaw the future and that if their motion was not the way it is, nothing here would be similar <to the way it is>.]

291a26 We have, then, said that the stars are spherical and that they do not move on their own.[518]

Just before this when he set out the view in a plausible way he said it[519] was spoken in an ingenious and clever way, but now he says it is spoken tunefully and musically, using these appropriate words ironically. And he next contests the view. He agrees with all the other things concerning proportion as being asserted in a genuinely musical way (for it is necessary that the sizes be proportional to the distances and the speeds of the motions to the sizes and – *if they come to be on their own* – the high and low pitches of the sounds to the speeds). But he censures the original hypothesis, according to which a sound incredible in its size is produced by their motion. He says that this is impossible, not just because it is strange that we hear nothing when such a great sound occurs, and that they do try to resolve the reason related to this (that is, the fact that we hear nothing), by proving that our hearing nothing is not a reason for <saying there is> strangeness, since there is another cause of our not hearing anything, habitualness. However, this does not provide a resolution. For if we did not perceive such great sounds because of habituation, it would be necessary that we not apprehend lesser ones as well; for the blacksmith does not hear a flying gnat either. And indeed, even things which we do not perceive because of habituation become perceptible when we pay attention.[520]

He says that what those people say is not just impossible for this reason, but also because they believe that only perceptive things which hear sounds are affected by them. For they assume that they escape absurdity by giving this explanation for why we do not hear. But they should have taken into account the fact that excessive sounds even shatter the bulks of bodies without soul which do not perceive; for the sound of thunder often splits stones and the strongest bodies. So if the sound of one clap of thunder does this much, then, if the bodies of the stars, which are so numerous and so large were to move, the sound passing through to things here in proportion to the magnitude in motion, a magnitude which is many times as great as the sound of the greatest thunder, must reach here, even if, as they say, we do not hear it because of familiarity. And consequently the strength of its force must be incredible enough to destroy everything here. It is reasonable, he says, <to conclude> that we do not hear <anything> and bodies here are not observed to be affected forcefully because there is no sound. For if there were a sound it would be necessary that we hear it and that things here be destroyed; but, if none of these things occur, it is clear that those things do not make a sound.

(291a6) He says that the reason why the stars do not make a sound in moving has become clear, and at the same time 'it gives witness that the things we have said are true' (and we said that the stars do not move[521] on their own). For what was a difficulty – why we do not hear <the stars> – and what made the Pythagoreans say that there are harmonies[522] in the things moving in heaven is evidence for us that the stars do not move on their own. For if they move – and this made the Pythagoreans say that there are harmonies of the moving things but we do not hear the sound – it is clear that they do not move on their own. For whatever moves on its own[523] makes a sound by making an impact, but whatever moves indirectly by being in something moving or being a part of what is moving, such as the parts in a ship, cannot make a sound, even if the ship does make a sound when it is moving through water which is still or moving in the contrary direction. But the ship itself does not make a sound if it is moving with the current, since then the boat is in a way a part <of the river> and in the water of the river as in a container. He says in the *Physics*[524] that the boat is as it were in place in the place of the whole river, since the place of things in a container is the place of the container; but the boat is in the water of the river as in a container.

Having clearly and gracefully transferred the difficulty to a ship and its parts, he adds, 'but if something moves in something which is not moving it makes a sound', as does a boat moving in still water or what moves through still air. It is impossible that what is in contact (for which he uses the word 'continuous'[525]) with what is moving and does not make an impact on its own, for example, the mast in a boat or the water in a clay jar or a ship in a flowing river, make a sound. And so here, he says, one should say – in accordance with the second hypothetical mode, which is also called the mode by means of three[526] – that if the bodies of the stars moved in a mass of air diffused up to the highest <region> or in a mass of fire – and fire is thought to be rather more appropriate because it rises up more than air – then 'it
467,1 would be necessary that they make a sound of enormous size'. But if the sound occurred, it would be necessary that it reach here and shatter things here. So, if what would follow necessarily from the stars moving on their own, namely that the sound reach here and shatter things here, is not observed to happen, it is clear that the antecedent would be false. Consequently the stars would not move on their own, either because of their own impulse, in the way things having soul do, or because of force, being projected by something or pushed or dragged. He says that this <situation> has come about as if nature foresaw the future and calculated in advance that if the stars did not move in the way they now do in the heavenly bodies by moving along with them, but rather moved on their own, 'nothing here would be similar' to the way it is now, but everything would be destroyed by the sound and rushing of the motion.

In connection with this passage Alexander correctly recognises that what is said here proves that it is reasonable that the spheres do not resonate when they move. For they do not divide by their motion any body through which they move and which remains still. Nor, indeed, is there any body of that kind in the heaven. Rather these spheres move internally, as Aristotle has said in several passages. They touch one another at their surface with no other body lying between them, their body not having any resistance or roughness or in general any extrusion.

Secondly he also correctly recognises that on the basis of what is said here Aristotle clearly believes that there is also providence[527] over things here, since he says that nature, foreseeing things here, made the spheres, but not the stars, move. For the efficient cause[528] also foresees this. And it is clear that Aristotle is now calling god nature, since what he calls nature in the strict sense – the starting point of motion and rest of what it is in – is not only not strong enough to have foresight, but it is such that it does not think or understand at all, nor does it perceive since it is naturally implanted first of all in bodies which do not perceive.[529]

Next Alexander tries to prove more directly through his own argument that the stars do not move on their own while the spheres remain still. He says:

> If they move in this way they either move (a) inside the sphere or (b) on the surface. But (a) if inside, either (i) they move through the heavenly body itself dividing the part of it in front of them or (ii) they move through some other generated body 468,1
> which is there or (iii) they move through a void which exists there.[530] But (iii) it is not possible for there to be a void anywhere else and much more impossible for there to be one in heaven. Nor (ii) is there some other destuctible body inside the divine eternal body, since that would be unnatural. So (i) if they moved through the heavenly body itself, cutting it, with the body flowing together again after being cut, the body, since it is divided, would be affected and, indeed, moist, since this is the nature of what is moist. So if the moist is opposite to the dry, and opposites have the same matter, it would be possible that this matter also take on the dry; but if this is so, the heavenly body would not be eternal or indestructible in its own nature.
>
> (b) But if they move lying on the surface of their own spheres and extruding from it and always divide the part of what lies alongside of the spheres which stands in front of them, this will either be a destructible body or a heavenly one, or they will move through a void. And again the same arguments can be given. But if there were some other body there, the divine body would no longer be spherical in itself nor would it be, as he said, turned out precisely.[531]

It is clear that the things said by Aristotle and[532] by Alexander are directed at those who say that the stars change place[533] on their own, and they do not create difficulty for anything in the Platonic doctrine that the stars also revolve in a circle about their own central points in the same way the whole spheres do. For they will not make a sound in being carried around with and in the spheres, nor will they do so while moving internally, just like the spheres, since the spheres do not make a sound on their own. The Pythagoreans said that a harmonised resonance is produced by the motion of the heavenly bodies and they inferred the motions scientifically from their proportionality to their distances, since the ratios of the distances of not just the sun, moon, Venus, and Mercury, but also of the other stars were apprehended by them.

I, indeed, wonder whether the explanation provided for why we do not hear in terms of familiarity and habituation is consistent with the Pythagorean story that Pythagoras at some time heard this harmony, despite the fact that it was familiar to him as much as to other people.[534] So, perhaps it is necessary to dissolve the objection in terms
469,1 of the philosophy of these men by saying that all things are not commensurable with one another and not all things, even in our world, are perceptible by everyone. Dogs who smell animals which are far away and which humans do not smell make this clear. By how much more then, in the case of things which are naturally so distant as indestructible things are from destructible ones and heavenly things are from earthly ones, is it true to say that the resonance of the divine bodies is not audible to mortal ears? And if anyone had this mortal body attached to its luminous[535] and heavenly vehicle and the senses in it purified[536] – whether because of a good fate or a good life or, in addition to these, because of ritual perfection[537] – he would see what is invisible to others and hear what is inaudible to others, just as is recounted of Pythagoras. And if a sound is produced by divine and immaterial bodies, it does not involve impact or scraping, but rather it activates the powers and activities of generative resonance and brings to completion the coordinate sense. And this sound has a certain analogy to the sound which accompanies the motion of mortal bodies, but it is a kind of impassive activity of the motion of the <stars>. Sound comes to be in our world because of the resonant nature of the air; so if there is no air which is acted on in heaven, it is clear that neither will sound be <there>. It seems that Pythagoras is said to have heard that harmony in the sense that he mentally grasped the harmonious ratios in numbers and what is audible in them and said that he heard their harmony.

Someone might reasonably raise the difficulty why we see the stars themselves with our visual sense organs, but we do not hear their resonance with our ears. We should perhaps[538] say that we do not even see the stars themselves; for we do not see their sizes nor their shapes nor their transcendent beauties nor even their motions on account of which there is a sound, but we see a certain sort of luminescence coming from them in the way that we see the light of the sun near the earth, but not the sun itself. But perhaps it would not be surprising if the visual sense, because it is less material and is constituted more by activity than by being affected and is much superior to the other senses, is deemed worthy of the radiance and luminescence of heavenly things, but the other senses are not suitably constituted for these things. But if someone can provide other, more plausible, explanations of these matters and matters like them, he will prevail as a friend, not an enemy.

But one should not pass over the way in which claps of thunder 470,1
burst stones and the strongest bodies. Now thunder is produced in two ways, either by a sudden breaking of denser clouds or by their colliding together. But in either way a blast of air is emitted with force from those clouds, and then this, moving with a rush and whirring sound, splits what it encounters. And the great power which blasts of air have is also made clear by gales, earthquakes, and the tides and floods of the sea.

(291a26) Finishing up his discussion of the stars, Aristotle says that he has said both that they are spherical – he will demonstrate this more precisely shortly[539] – 'and that they do not move on their own'.[540] He obviously means their observed change of place[541] about which he has been giving demonstrations all along. And in addition it is therefore clear that he has demonstrated that they do not make a sound, since this is also in itself appropriate to the present subject and since he infers their lack of motion from their not making a sound. The fact that the Pythagoreans, as he says (or, if he doesn't mean them, I don't know who – since he doesn't say this clearly about the Pythagoreans), think it is necessary that a sound be produced when such large bodies move[542] has played a role in his inference that if they do not make a sound they do not move. For if it is necessary that if they move they make a sound, then if they do not make a sound, they do not move. But it is clear that that has been added as what follows for those who say that in moving they make a sound. If, however, someone were to say that they do not change place <on their own> or because of bodies which make a sound, it would no longer be true to say that if they do not make a sound they do not move, since Aristotle too says that the spheres move, but denies that they make a sound and does not make their not making a sound evidence that they do not move (because they move internally but do not move through some other body). And so too, if the stars move internally in the same way as the spheres, they are not constrained to make a sound; nor if they do not make a sound do they thereby not move.

Notes

1. cf. the prologue to the commentary on the first book at 1,1-6,27, especially 4,25-34.

2. i.e. Plato in the *Timaeus*.

3. *hê en tautôi monê*, also at 374,6.

4. In the lemmas I have translated the whole passage discussed by Simplicius. In the lemmas in Heiberg's text only the first and last few words are given. So in this case one has ' "That the whole heaven has not come to be" up to "to the belief that the cosmos is immortal and eternal." ' On the text of the lemmas (and of some of Simplicius' citations and paraphrases of Aristotle) see the introduction to Moraux's edition, pp. clxxxv-clxxxvi.

5. On the Aristotelian commentator Alexander of Aphrodisias, to whom Simplicius constantly refers in this commentary, see Sharples (1987).

6. cf. the prologue to the commentary on the first book, especially 5,13-7.

7. cf. Aristotle at 1.10, 279b12-7. For some of the issues of interpretation raised by these remarks see, for example, Long (1974) and Wright (1981), pp. 40-8, for Empedocles, and Kirk (1954), pp. 306-65, and Kahn (1979), pp. 132-52, for Heraclitus.

8. See the next lemma at 370,1.

9. That is, of timelessness.

10. Simplicius practically quotes *Timaeus* 38B6.

11. *Timaeus* 37D3-7.

12. For examples of such people ('Plutarch, Atticus, and many others') see Proclus, *in Tim.* (Diehl (1903-1906)) vol. 1, 276,30-277,7, with Festugière's (1966-1968) note at 276,31.

13. cf. *Phaedo* 106B1-D4, where Socrates argues that the soul is both immortal and indestructible. See the two commentaries of Damascius ad loc. (Westerink (1977)), from which I quote one sentence in Westerink's translation (Damascius II, 76): 'Socrates includes forms as well as things to which a certain form is essential to make sure that soul will fall under his assumptions, no matter whether one takes it to be life pure and simple (as simple forms, the hot itself and the cold itself) or as a substance to which life is essential (as fire, which has heat as an essential quality)'.

14. *Timaeus* 41B.

15. Simplicius takes up these words at 371,16.

16. Starting at 371,16 Simplicius discusses two interpretations of these words: Alexander's, according to which they mean that limits are among things which contain; and his own, according to which they mean that being a limit belongs to (i.e. is a property of) things which contain.

17. *athanaton* misprinted as *athananton* in Heiberg.

18. *Cael.* 1.3, 270b5-9.

19. Reading *diabeblêmenoi*, following the suggestion of Praechter (1924).

Heiberg prints the *diablepomenoi* of F. A has *de blepomenoi*, Karsten *anablepomenoi*. On this paragraph attacking the Christians see Hoffmann (1987), pp. 69-70.

20. Hoffmann (1987), p. 70, cites Matthew 5:35.

21. Simplicius apparently recalls fr. 96 of Heraclitus on corpses (DK22B96; for numerous citations of the fragment see Marcovich (1967), pp. 407-10). I quote Hoffmann (1987), p. 70: 'This remarkable extract reflects all that pagan piety saw as most revolting and shocking in the veneration of Christ's dead body and of the relics of martyrs'.

22. I have tried to translate the *genomenên* of A. Hoffmann (1987), p. 70, translates the *genomenon* of F, printed by Heiberg, as 'as if <the heaven> was created simply to give free passage to their insolence'.

23. i.e. when there are no valid syllogisms with two affirmative premisses in the second figure.

24. *tas* misprinted as *tan* in Heiberg.

25. The *aphôrismenon* of F and Karsten seems preferable to the *aphôrismena* printed by Heiberg.

26. *Timaeus* 41D2-3.

27. *sunkraseôn*. I thank David Furley for clarifying the meaning of these words.

28. *Cael.* 1.3, 270a25-35. Simplicius' discussion of that passage (111,3-115,20) is very useful for understanding what he says here. It is important to realise that Simplicius is willing to allow some kind of qualitative change (*alloiôsis*) in the heaven as long as it isn't the kind which involves 'affection' (*pathos*); see e.g. 114,4-6.

29. On the *hupekkauma*, a highly flammable vapour occupying the highest sublunary region, see *Meteor.* 1.3-4 and Simplicius' commentary on Chapter 7, starting at 438,28. In both the *Meteorology* and Chapter 7 Aristotle says that the heat and light which we associate with the heaven are not to be found there but are to be explained in terms of the way in which the motion of the heaven affects what is beneath it (the word *hupekkauma* is not used in *De Caelo*). At 2.7, 289a31-3 and *Meteor.* 1.3, 341a12-30 Aristotle associates the production of heat especially with the sun. His doing so leads to the difficulty which Simplicius raises here: why, when the moon is below the sun, should the sun heat up what is below the moon and not the moon itself?

30. This is Alexander; see 440,23 ff. and especially Alexander's discussion at *in Meteor.* (*CAG*, vol. 3.2) 18,14 ff., where he gives the example of electric eels in nets.

31. As the words 'vital and perfective' suggest, in this paragraph Simplicius is expressing his own view, which he thinks of as importantly different from Alexander's. See below 436,4-437,29; see also the discussion in Simplicius' commentary on the first book at 111,3-115,20.

32. That is, it is spatially extended.

33. See above 366,25 ff.

34. i.e. the heaven. For Simplicius the upper region is what lies just below the heaven.

35. Most of this sentence is added to DK31A49; see DK, vol. 1, p. 499.

36. On Ixion see below 377,2 ff.

37. *manteia*; for Simplicius' understanding of this word see below 382,28-32.

38. The lemma has *ta nun*; Moraux prints *to nun*. Cf. Simplicius' paraphrase at 382,20 with the note.

39. 284a14 at the end of the previous lemma.

40. Actually line 517 of Hesiod's *Theogony* (West (1966)).

41. *Odyssey* (van Thiel (1991)) 1.53-4.

42. For discussion of Empedocles on this issue see O'Brien (1969), pp. 51, 262-5, 298-9. Simplicius' addition of Anaxagoras seems somewhat difficult to square with Aristotle's statement (2.13, 294b13-14) that for Anaxagoras and others the earth remains fixed because of its flatness. Most of the discussions of the vortex in Anaxagoras refer to its role in separating things and in raising the heavenly bodies aloft (for the sources see Gershenson and Greenberg (1964), pp. 348-9). Simplicius repeats the statement about Anaxagoras just below at 375,25-8 and in his commentary on Chapter 13 at 526,34-527,1. He repeats Aristotle's statement about flatness at 520,26-31, and finally tries to harmonise the two accounts of Anaxagoras at 527,31-5.

43. At *Timaeus* 39E10 Timaeus describes the kinds of living things: 'There are four of them, one the heavenly genus of gods, another the winged species which flies through the air, third the aquatic species, and fourth the one which walks on the ground. <The demiurge> fashioned the greatest *idea* of the divine <kind of living thing> from fire so that it would be the most beautiful and the brightest to see.' When Alexander refers below (378,8-9) to 'those for whom the body which moves in a circle is either one of the four <elements> or composed out of them' he refers to an interpretation of Plato based on this passage. Alexander (and apparently Aristotle) took this position on the composition of the heavenly bodies to mean that they could not have a natural circular motion so that the motion of heaven which Plato ascribes to soul would have to be forced. According to Simplicius, Plato thought the heaven was composed of the purest forms of the four elements and predominantly of fire (see 435,33-436,3). For a description of Simplicius' attempt to reconcile this interpretation of Platonic doctrine with Aristotle's apparent belief in a fifth element see Hoffmann (1987), pp. 76-9.

44. *Timaeus* 36E2-5, which, however, is longer: 'Extended in every direction from the middle to the last heaven, enveloping the heaven from outside in a circle and itself turning in a circle in itself, it made a divine beginning on an unceasing and rational life for all time.'

45. What follows is based on Orphic materials, for which see West (1983) and particularly pp. 75, 140-1. The most closely related extant text is Proclus, *in Tim.* (Diehl (1903-1906)) vol. 1, 173,1-6 (= Kern (1922), no. 215): 'The theologians say that after the dismemberment of Dionysos (which indicates the dividing progression from the undivided demiurgic activity into the universe caused by Zeus) the other Titans were allotted other portions, but Atlas was established in the western regions and holds up the heaven.' For further discussion see the note on this Proclus passage in Festugière (1966-1968).

46. Oracle 6 Des Places (1971), which can be consulted for text and interpretation; see also Majercik (1989). Des Places prints: *hôs gar hupezôkôs tis humên noeros diakrinei, | pur prôton kai pur heteron speudonta migênunai.* Heiberg prints only the second line as a quotation. Des Places prints the two lines with *humên* from Damascius *in Parm.* (Ruelle (1889)) 2,131,29 substituted for the *ou mên* printed by Heiberg following A (*hêmin* in F and Karsten).

47. For the addition of Anaxagoras see the note on 374,34. For Democritus compare the description of Leucippus' cosmogony in Diogenes Laertius (Marcovich (1999)) 9.30-3 (= DK67A1); see also Guthrie (1962-1981), vol. 2, pp. 404-13, 422-4.

48. There is parallel material (on Democritus) at 262,9-15 of Philoponus *in Phys.* (*CAG*, vol. 16).

49. *emphronos*, the same word which I have translated 'mental' at 284a32 in the lemma and in the present discussion.

50. *Timaeus* 36E4-5.

51. For the full text of the passage see the note on 375,3.

52. That is, Aristotle represents the statement of the soul being extended through the cosmos in terms of Ixion being stretched out on his wheel. Cf. 378,32-379,4.

53. Our earliest source for the myth is Pindar's second Pythian Ode. For some discussion see Gantz (1993), pp. 718-21.

54. In this story, the father of the centaurs.

55. I have tried to translate Karsten's *summikton phusin logikôn te kai alogôn energeiôn*. Heiberg prints *surphetên logikôn te kai alogôn energeian*. *surphetên* is his conjecture for the *surphusin* of A. F also has *summikton phusin*, but, apparently, only Karsten has *energeiôn*.

56. 377,12-20 are fr. 230 of Kern (1922). Kern prints *anapsukhai* in agreement with F and Karsten where Heiberg prints the *ampsukhai* of A. I have followed Kern.

57. A strong statement of the harmony of Plato, Aristotle, and divine truth. On this conception in Simplicius see Hoffman (1987), pp. 77-9, and for a more general discussion Hadot (1978), pp. 67-76.

58. *Cael.* 2.2, 285a29-30.

59. Heiberg here prints *monon autôn* with F and Karsten, against the *monôn autôn* of A. See the first note on the lemma at 482,1 in the commentary on Chapter 12.

60. The manuscripts of Aristotle have *metekhontôn*, as Simplicius also does at 78,28 in the commentary on 1.3. But here he has *metakhontôn autôn*: see also the note on 388,24.

61. *Cael.* 2.12, 292a18-21.

62. Reading *sômatôn* rather than the *sômatos* printed by Heiberg.

63. Simplicius thinks Alexander is stuck with the view he wants to assign to Plato.

64. Aristotle's general definition of soul in *DA* 2.1. Simplicius makes the same point about Alexander at *in Phys.* (*CAG*, vol. 10) 1219,10-11. For a survey with bibliography of Alexander's known views on the soul and intellect see Sharples (1987), pp. 1202-14.

65. cf. 380,16-9.

66. See the note on 375,3.

67. cf. Simplicius (?), *in DA* 40,22-6.

68. I have punctuated as follows: period after *sômatos* in 378,37, comma after *pariegagen* in 379,1, comma after *enomizen* in 379,1 and no comma after *Platôn*.

69. A close paraphrase of *Timaeus* 36D9-E1.

70. See the note on 375,2.

71. Circular, forward, backward, up, down, right, left.

72. *Timaeus* 34A1-4.

73. Rose (1886), p. 164, and others take this to be a reference to a work called *Excerpts from the 'Timaeus' and the Works of Archytas* included in a list of Aristotle's works in Diogenes Laertius ((Marcovich (1999)) 5.25); cf. 296,16-8 in the commentary on Book 1. This last sentence apparently means that Simplicius does not think Aristotle is attacking Plato here, but he does not tell us whom Aristotle is talking about.

74. Reading *autois*, a suggestion of Heiberg, who prints *tois*. Karsten has *autois tois*, which is also a correction of *autois* in F.

75. A city in Lydia or perhaps Caria. All that we know of Julian comes from the present passage. Given the location of Tralles and the fact that the other person mentioned by Alexander here is his own teacher Herminus, it would seem likely that Julian was associated with the same Peripatetic school. On Herminus, who is mainly known for work on the *Organon*, see Moraux (1984), pp. 361-97. On the question of the role of soul in the motion of the cosmos see also Merlan (1943), pp. 179-82, and Sharples (1983).

76. That is, east-west.

77. *to ep' apeiron kai aei kata ta auta kai hôsautôs kai peri ta auta kai en tôi autôi kai hosa toiauta.*

78. e.g. in *DA* 3.4-5.

79. Reading *tên dunamin tês psukhês* with F and Karsten instead of the *tên dunamin tên psukhên* printed by Heiberg, following the first hand of A; in the second hand the *tên* is corrected to *tês*, but *psukhên* is left. b and a marginal notes in K suggest a reading of *tên dunamin tên kata tên psukhên*. At *in Phys.* (*CAG*, vol. 10) 1219,3-7 Simplicius quotes most of this passage, but not this sentence and not the words 'in accordance with which nature and soul the body moves in a circle'. The only significant difference between Heiberg's text here and Diels' in *in Phys.* concerns the words 'for soul is more complete than nature' (*teleiotera gar phuseôs hê psukhê*), which in Diels' text are 'soul is a more complete nature' (*teleiotera gar phusis hê psukhê*); Diels' text seems to me more likely to be what Alexander wrote.

80. What Alexander says here should be compared with parallel statements in his work on the cosmos (Genequand (2001), 4-22), and in his *Quaest.* (1.25; Bruns (1892), 39,8-41,19, especially 40,23-30). On the latter see also Sharples (1992) ad loc., and Bodnar (1997).

81. cf. *Phys.* 2.1, 192b21-4.

82. Simplicius summarises the upshot of *DA* 2.1.

83. *Cael.* 2.12, 292a18-21.

84. Simplicius returns to this topic briefly at 387,5-19. He discusses Philoponus' view on the question at 78,12-80,23 in the commentary on Book 1.

85. Simplicius takes issue with the doctrine that elements only have their nature when they are in their natural place. Cf. the commentary on Book 1, at 21,33-23,10.

86. *Cael.* 2.2, 284b33-4. Simplicius here has *allôn* where our texts of Aristotle have *apsukhôn*; but Simplicius has *apsukhôn* at 387,5 below and at 288,3 of *in Phys.* (*CAG*, vol. 9). It seems likely that, if Simplicius did not write *apsukhôn* here, he expected his audience to supply it.

87. *Cael.* 2.2, 285a28-9.

88. *Laws* 894C2-3. Plato does not explicitly call this motion nature, but the label is reasonable enough in the context.

89. Heiberg prints the *ekeinos* of A, noting that b has 'Plato', F and Karsten 'Plotinus'. F and Karsten also have *dia ti*, which I have translated – Heiberg prints *dioti*. With *dia ti* what follows is a virtual quotation of Plotinus, 2.2.1.1. For other citations of these words in antiquity see the note on 2.2.1.1 in Henry and Schwyzer (1951). Merlan (1935), pp. 416-18, drew attention to Heiberg's mistaken assumption that the divine man was Plato, and in the same article pointed out the close parallelism between this paragraph and a passage in [Alexander]'s commentary on the *Metaphysics* (*CAG*, vol. 1, 706,31-707,11).

90. See the last note on the lemma. Here Heiberg prints *to nun* while noting that F and Karsten have *ta*.

91. For the reference see 384,10-21 below.

92. Simplicius 'explains' this word at 383,8-10. For a minor textual issue (the order of the words 'right' and 'left') see Moraux (1954), p. 174.

93. *diastaseis*, sometimes used for the three pairs just mentioned.

94. Simplicius explains this term at 386,31-387,4.

95. On this remark see 387,5-19.

96. cf. 387,19-24.

97. Moraux brackets these words; they do not occur in Simplicius' paraphrase at 384,4-6.

98. I have preserved the distinction between these two phrases in my translation, but Aristotle appears to abandon the alleged distinction at 285b16-8, and Simplicius follows him.

99. Simplicius' formulation here is based on Plato's *Cratylus* 397C8-D6, where Socrates connects the word *theoi* (gods) with the verb *thein* (to run).

100. For the ten pairs see Aristotle, *Metaph.* 1.5, 986a22-6 with the note in Ross (1953).

101. *IA* 4, 705a27-31.

102. *IA* 4, 705b9-10. Aristotle has *to opisthen*, as do F and Karsten. Heiberg prints *opisthen*.

103. The *kai* printed by Heiberg is not in Aristotle (or in Karsten).

104. *IA* 4, 705b13-18.

105. A later Greek term, not in Aristotle, defined in the *OED* as follows: 'A general name for various animals of low organisation, formerly classed as intermediate between animals and plants, being usually fixed [i.e. stationary] and often having a branched or radiating structure, thus resembling plants or flowers: as crinoids, hornwracks, sea-anemones, corals, hydroids, sponges, etc.'

106. That is, allegedly, if there are surfaces, there are lines, but there could be lines and no surfaces.

107. Aristotle gives senses of priority in *Metaph.* 5.11 and *Cat.* 12. Simplicius more or less follows the *Metaphysics* list. See the notes which follow.

108. Simplicius may take the notion of priority in position from *Phys.* 4.11, 219a14-6, but the idea, although not the term, is also at *Metaph.* 5.11, 1018b9-14.

109. *Metaph.* 5.11, 1018b26-9 and *Cat.* 12, 14a34-b3 (from which Simplicius' example comes).

110. *Metaph.* 5.11, 1018b22-6; the term 'honour' comes from *Cat.* 12, 14b3-8.

111. *Metaph.* 5.11, 1019a1-11.

112. *Metaph.* 5.11, 1018b14-19; for the equivalence of priority in time and priority in coming to be see *Metaph.* 9.8, 1049b17-1050a3.

113. This brief (but reasonable) objection comes in rather curiously here. The fact that the MSS go on to talk about a 'third' objection (see the next note) suggests that something may have gone wrong with the text.

114. Reading *pempton* in place of *triton*. Simplicius has already spoken of four criticisms. He is now apparently at 285a26 and considering taking the words 'leaving out more important principles' as summarising all the material in 285a10-25 as a single objection and the words 'thinking that the principles exist equally in all things' as a second objection. The content of this second objection is, of course, provided by 284b30-285a10.

115. Not translating Simplicius' resumptive *triton d'oun autois ê deuteron epagei*.

116. References to the right nature of a person are fairly common; see e.g. Olympiodorus' commentary on the first *Alcibiades* (Westerink (1956)), 90,5; although right and left were always associated with good and bad fortune respectively, my best example of right fortune is a reference to a person's rightness of fortune in the *Suda* (Adler (1928-1938), pt 1, 20,22).

117. That is, some things with up/down and front/back do not have right/left.

118. Most of 386,9-23 constitutes Aristotle fr. 200 Rose (1886).

119. The *panta* printed by Moraux is missing in Heiberg, who notes that it occurs in F and Karsten.

120. Heiberg prints *de* where Aristotle has *gar*. Again F and Karsten agree with Aristotle.

121. Heiberg does not print the *pasas* which is in Aristotle and in F and Karsten.

122. Heiberg's reference to *Phys.* 8.4 seems plausible.

123. cf. 380,29 ff.

124. For Simplicius, as for Neoplatonists generally, the World Soul is not 'in' the world; see, for example, Wallis (1995), pp. 51-3.

125. At 284b30-285a10 in the preceding lemma.

126. What follows is much more clearly Simplicius than Aristotle; see 385,7-29.

127. *dianoumetha* comes after *pampan* in Aristotle, which is where Simplicius has it at 378,15. But here he has it after *monadôn.*

128. *monon autôn*; see the first note on the lemma at 482,1 in the commentary on Chapter 12.

129. *Cael.* 2.12, 292a18-21.

130. Presumably Simplicius is referring to *Nicomachean Ethics* 6.2, as Richard Kraut pointed out to me.

131. On this 'also' see 389,11.

132. Heiberg: *homoiôn ge ontôn pantôn tôn moriôn.* The MSS of Aristotle and Karsten have *homoiôn g' ontôn tôn moriôn apantôn.*

133. cf. 384,27-385,2.

134. Heiberg omits a *pôs* which is in our texts of Plato and is printed by Karsten. Simplicius also omits the words 'as most of the people of this kind are bold enough to assert'.

135. *Laws* 895A6-B1.

136. Simplicius offers an explanation of this remark at 390,21-5.

137. Simplicius explains these words at 391,30-392,5. Circular motion to the right (*epi ta dekhia*) is what we would call counterclockwise; see Braunlich (1936).

138. On the remainder of this paragraph see 392,12-32. 'Towards the right' translates *pros ta dekhia.*

139. *Phaenomena* (Kidd (1997)) ll. 24-6. According to a scholiast, 'Ocean' means 'the horizon'.

140. The notion of having one's back on the axis and head towards the south does not specify the position of the person imagined. The easiest way to make the description specific is to say that the east or place of rising is to the person's right; but then the description becomes circular.

141. *Iliad* (West (1998)) 12.239-40, Hector speaking, dismissing the appearance of birds as omens. Heiberg prints *dekhia esti*, following C, West has *dekhi' iôsi*, as do F and Karsten. A has only *dekhia.*

142. Simplicius omits *merei* in what is otherwise a citation.

143. 392,16-32 is Aristotle fr. 205 Rose (1886).

144. Here an argument of the form 'P or Q or R; not P; not Q; therefore, R'.

145. This word (*elaphrôs*), which doesn't seem to have much point, is deleted in K (perhaps by Bessarion) and omitted by Karsten.

146. That is, having two rights.

147. cf. 392,21-4.

148. cf. 392,12-32.

149. In the next chapter.

150. In 1.4. Simplicius' point is that Aristotle 'really' means 'reverse' (*anapalin*) not 'opposite' (*enantios*).

151. See the beginning of the next chapter.

152. Translating the *tên planômenên sphairan* of F and Karsten rather than the *tên <tôn> planômenôn sphairan* printed by Heiberg. (A has *tên planômenôn sphairan.*)

153. Reading *opsetai* with Karsten (b has *videbit*) rather than the *opsei* of F printed by Heiberg.

154. Chapter 4.

155. Chapter 5.

156. See 419,7-421,37.

157. Karsten does not print this lemma, which is, in fact, included in the scope of the preceding lemma as printed by Heiberg.

158. The remainder of the chapter is a single argument, which is summarised by Simplicius at 403,27-35.

159. See 366,7-8 in the outline of Book 2 at the beginning of the commentary.

160. See 394,1-27.

161. In the previous chapter at 285b14-27.

162. One should read *diaphoron* with F and Karsten rather than the *diaphora* printed by Heiberg.

163. In 1.4.

164. *andrizomenos*, surely sarcasm.

165. *sumbebêkota*, standardly translated as 'accidents'. Simplicius expects demonstrations to be based on *kath' hauta huparkhonta*, and supposes that is what Aristotle means here.

166. *Timaeus* 29B3-D2.

167. I take Simplicius' point to be that we would have absolutely no perceptual awareness of something on the ground ten stades away from us as we do of something visible to us although the distances between the two objects stand in some ratio.

168. For the lemma Heiberg prints *hekaston de* with A, where Moraux and Karsten print *hekaston estin*. In the paraphrase/citation at 396,29-30 Simplicius has simply *hekaston*, but the *estin* is replaced by a subsequent *einai*. If I understand Heiberg's apparatus for 396,26-30, the original text of F had as lemma at least the first part of what is printed by Moraux, but it was replaced in the margin with the text printed by Heiberg.

169. *energeia*; the word translated as 'function' is *ergon*, a word which could also mean 'product' (a word which here translates *apotelesma*). Simplicius (see 396,30) treats them as synonymous, and ends up substituting *energeia* for *ergon* (see 397,4). His running these things together produces a certain obscurity in his discussion. He seems to hold that every art and science is for the sake of its activity or exercise, some arts are *also* for the sake of their product; see 397,32-398,9.

170. Simplicius is willing to allow the heaven to be called a god, although in his view Aristotle's 'real' god is the prime mover of *Metaph.* 12. Alexander has a brief discussion of this issue in *Quaest.* 2.6 (Bruns (1892)).

171. cf. *Phys.* 8.8.

172. Simplicius raises questions about each of the first two assertions in the lemma.

173. *Hexeis.*

174. cf. *Cael.* 1.9, 279a18-22.

175. On the fourth-century Neoplatonist Syrianus see PW vol. 4A, cols 1728-75.

176. In Chapters 13 and 14.

177. cf. 401,5-8.

178. That is, the word 'heaven' in 286a13 refers to the whole world. See below 398,25-399,4. For criticism of Simplicius' reading of this passage, see Federspiel (1995), pp. 511-12.

179. In Chapters 1-4.

180. Simplicius' reasoning here is not clear to me, but, then, neither is the Aristotelian text he is trying to understand. It does seem clear that Aristotle wants the stable thing in the centre about which the cosmos or the heaven rotates to be the earth, not a point, but he has not given any reason why the earth couldn't rotate about a point. Simplicius says, what is incorrect from a modern perspective, that the central point of a revolving sphere moves with the sphere, and adds that the central point of the cosmos does not change place.

181. Here again Simplicius is relying on Chapters 1-4 of *On the Motion of Animals.*

182. Simplicius, then, understands Aristotle to be saying, 'why doesn't the whole cosmos (body of the heaven) revolve in one direction? Because some body which is a part of the whole cosmos remains fixed at the centre. But no part of heaven (it) is able to remain fixed anywhere'.

183. Usually known as Nikolaos of Damascus, as he is called by Simplicius, *in Phys.* (*CAG*, vol. 9) 23,14 and 149,8 and *in Epict.* (Hadot (1996)) 37.16. Simplicius refers to him as Nikolaos the Peripatetic at the beginning of the commentary on Book 1 (3,28). Fragments of the work to which Simplicius refers are preserved in Syriac and available in Drossart Lulofs (1965). See also Moraux (1973), pp. 445-514.

184. That is to say, there might always be unnatural things like monsters, but there would not be one monster which lasts for all time; on the other hand, the heaven is a single natural eternal body.

185. cf. *Cael.* 1.3, 269b20-6.

186. In Chapter 14.

187. Heiberg prints *All' ei*, Moraux and Karsten *Alla mên ei.*

188. Reading *kinêton* for the *akinêton* printed by Heiberg and Karsten; cf. e.g. 400,20: 'for fire is always moving'. Perhaps one should emend to *aei kinêton.*

189. Not in this passage, but see *GC* 1.3, 318b14-8 with the note ad loc. in Joachim (1922).

190. For a revision of this formulation see the next paragraph.

191. Heiberg refers to Chapter 5, in which Aristotle explains the sense in which the matter of the elements is one; Chapter 4 is perhaps more likely, since there Aristotle describes the relative weight and lightness of the elements. So too Leggatt and Moraux.

192. *GC* 2.3-4, where Aristotle characterises the elements in terms of hot/cold, wet/dry.

193. Heiberg prints *diakosmia tade*, following A; Karsten *ta enkosmia tessara.* I have translated the *diakosmia ta tessara* of F, which Heiberg thinks may be right. So apparently did Bessarion since he corrected K to *ta enkosmia* d^a. For the Greek see Heiberg's apparatus to 401,9.

194. Simplicius paraphrases *Timaeus* 31B4-32C4.

195. Simplicius discusses this difficult pair of sentences from 401,21-402,16.

196. Simplicius (402,37) does not provide a reference; modern commentators refer to *GC* 2.10.

197. In Chapter 8.

198. That is, intelligible things have an 'eternal life', but they do not move in the relevant sense.

199. *kinêtos*.

200. Heiberg prints *hou mê endekhetai kata phusin tên kinêsin einai aidion*. Moraux (and Karsten) prints *hou mê endekhetai einai kata phusin tên kinêsin aidion*.

201. This commentary on 286a31-b2 may need explanation. In the first sentence Aristotle says that there must be coming to be because none of the elements is eternal. In the second and third Aristotle gives an (obscure) argument that the elements are not eternal. And in the fourth he reaffirms that there is coming to be. Simplicius first gives Alexander's interpretation. Alexander appears to have taken the second sentence as something like:

1. A moving thing is eternal if and only if its natural motion is eternal.

It is easy to construct out of the *Physics* and this assumption the conclusion that:

2a. The elements are not eternal,

and the irrelevant conclusion, apparently drawn by Alexander:

2b. The heaven is eternal.

The question remains what the point of the third sentence is. Alexander took it to be elliptical:

3. (i) The elements move, <(ii) but their motion is not eternal>.

The need for this addition is accepted by many scholars; see, e.g. Guthrie ad loc. But Alexander added an explanation of why Aristotle asserted what is apparently obvious, namely, that the elements move. His explanation is that he means:

3′. Although the elements do not have an eternal motion, they are not unmoving (and for that reason eternal).

Simplicius is perfectly happy with 3′, but he thinks that there is a problem with Alexander's 1 because Alexander substitutes 'moving' (*kinoumenos*) for Aristotle's 'movable' (*kinêtos*). 1 should read:

1_S. A movable thing is eternal if and only if its natural motion is eternal.

And Simplicius suggests that what is eternally in motion is not in the strict sense 'movable'; only things which sometimes move and sometimes rest are. This means that only 2a and not the irrelevant 2b can be inferred from 1 and the *Physics*, and the third sentence simply asserts the relevant doctrine of the *Physics*:

3_S. The elements are movable, that is they move sometimes and rest sometimes.

This enables him to also have 3′.

202. Simplicius now 'unpacks' 286b8-9. In the next paragraph he offers an explanation of Aristotle's 'brief' formulation.

203. That is, *MA*, Chapters 1-4; cf. 398,18-20.

204. *Phys.* 8.1-2.

205. Chapter 3.

206. Simplicius here recalls the beginning of the chapter, 286a8-12.

207. i.e. the Stoics. For the doctrine of cosmic conflagration (*ekpurôsis*) see e.g. Long and Sedley (1987), vol. 1, pp. 274-9.

208. Heiberg prints *tôn zôiôn huparkhis* without the preceding *hê* of F and Karsten. The *hê* seems necessary.

209. Roughly a ruler without markings and thus useless for accurate measurement. Cf. Socrates' description of himself at *Charmides* 154B9 with the scholium in Greene (1948), pp. 113-14; the proverb was apparently 'a white measure for a white stone'.

210. Since the 'second resolution' completely undercuts the very statement of the difficulty and the 'first resolution' concedes to the Stoics much more than one would expect Alexander to concede, it seems more likely that in the first resolution Alexander was only pursuing a line of argument (however repugnant the line might be to Simplicius).

211. At 409,2-3 Simplicius cites these last words without 'first', giving the sense 'most appropriate for both its substance and its nature'. His paraphrase of the argument at 406,3-8 does not make clear how he reads the text, since he refers to sphericity as the naturally first shape and the heaven as the naturally first body, and he does not mention appropriateness. I am inclined to think Moraux is right not to count the 'citation' at 409,2-3 as evidence for Simplicius' text.

212. Simplicius tacitly corrects Aristotle's assertion that all curvilinear figures are contained by one line.

213. Simplicius gives no reference here, although he may have in mind *Cael.* 1.2, 269a19-23. There are closer parallels at *Phys.* 3.6, 207a7-9, and *Metaph.* 5.16, 1021b12-13 and 10.4, 1055a12.

214. Reading *dêlon oun*, following a suggestion of Heiberg, instead of the simple *dêlon* which he prints.

215. cf. 385,15-24 with the notes. At *Metaph.* 7.3, 1028a31-b8 Aristotle argues in an unclear way that substance is prior in *logos*, knowledge, and time, but that passage does not seem to have any bearing on what Simplicius says here. His idea of priority in substance is perhaps best understood in terms of completeness, where mature adults are more complete than creatures in earlier states and 'eternal causes' (Forms) are more complete than their instances. Simplicius here argues that what is complete (in this sense) is prior to what is incomplete, a point which does not seem terribly relevant to the passage he is commenting on.

216. Reading *to te gennêsan* with Karsten and Bessarion's correction of K rather than the *to te genêton* printed by Heiberg. Heiberg reports that A and F have *to te gennêton*, b *id quod genuit*.

217. There is no question that Aristotle is referring to the construction of (four of) the regular solids out of triangles in the *Timaeus* (53C ff.). Simplicius perhaps uses the nebulous and even misleading *phusikoi* because of the complexity of his explanation of the Platonic procedure, on which see Mueller (2000). The point he wants to insist on here is that – on his interpretation – the division of the regular solids into triangles composing their surfaces is not a division into parts in the ordinary sense.

218. Simplicius gives a reason for correlating one with the circle, but he does not provide any clue about whom, if anyone, Aristotle is referring to.

219. Aristotle appeared to have switched from circles to spheres at 286b23, but at 286b33-287a2 he appears to return to circles. Simplicius smooths this over by invoking the comparison of circles and spheres at 286b25-6.

220. In *Cael.* 1.2, 269a18-32. The charge of circularity was made by Philoponus. See Simplicius on Book 1 at 45,2-27, the upshot of which Simplicius summarises in the last sentence of this paragraph.

221. Simplicius avoids Aristotle's loose use of the term 'continuous' (*sunechês*). Cf. 466,29 with the note.

222. See *Phys.* 4.6-9.

223. See 287a30-b4 with Simplicius' discussion starting at 414,20. His point is that, even if the last argument shows that the outer surface of the fire under the moon is spherical, it does not show that its inner surface is; for that one would need to know that the outer surface of what is under fire is spherical.

224. The *ei hê* of F and Karsten seems preferable to the *ei* of A printed by Heiberg.

225. Simplicius omits a *zôia* in our texts of Plato.

226. Simplicius has *hosa*; Rivaud (*Oeuv. Comp.*, vol. 10) prints *hoposa*.

227. Simplicius has *etornôsato* where our texts of Plato have *etorneusato*.

228. Simplicius omits a *tas* in our texts of Plato, but he has it at 418,9.

229. *Timaeus* 33B1-7.

230. See the note on 286b11 in the lemma.

231. Cf. 408,21-6. Now Simplicius perhaps has in mind the point he mentions at 418,8-10 at the end of the discussion of the next lemma.

232. *Periphoras*. Moraux prints *phoras* (motion). But Simplicius appears to cite the lines with *periphoras* at 409,30-2 (where Karsten prints *phoras*), although in a virtual quotation at 410,11 he has *phoras*.

233. See *Cael.* 1.2, 269a18-32.

234. See *Cael.* 1.9, 279a11-18.

235. The difficulty is to see in what sense there would be body outside, say, a cubical cosmos. Simplicius attempts to explain this in a tortuous sentence at ll. 25-7, the point of which seems to be that in the revolution of a rectilinear figure part of the figure would come to occupy a place which had been outside it.

236. i.e. the opposite surfaces of the 'lentil'. The contrast Simplicius goes on to make can be understood by thinking of a wheel revolving around an axle ('poles lying in the centre of the planes') or around one of its spokes ('poles lying in the periphery').

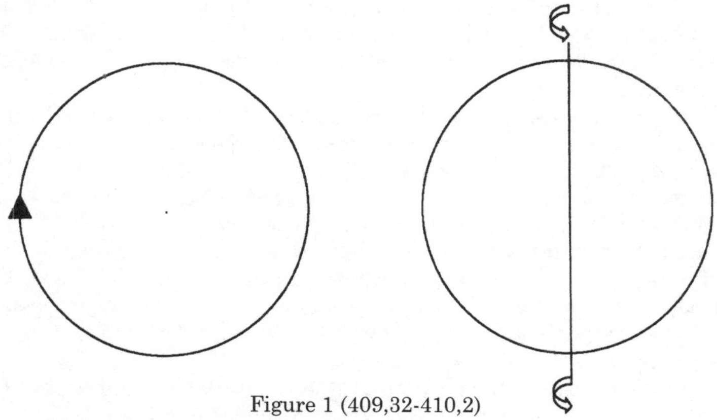

Figure 1 (409,32-410,2)

237. In this case Simplicius does not mention that if the poles determined a straight line perpendicular to the axis, the heaven would not occupy the same place.

238. The parallel passage in Ptolemy (Heiberg (1898-), vol. 1.1, 12,19-13,3) occurs in a chapter entitled 'That the heaven moves like a sphere': 'Speaking in summary, if someone were to hypothesise a shape other than spherical for the motion of the heavenly <bodies>, it would be necessary that their distances from the earth upward would become unequal, wherever and however the earth was assumed to be, so that in each revolution the sizes of the stars and their distances from one another should appear different to the same people, since at one time they would be at a greater distance and at another time less. But this is not seen to happen.' It appears from Theon's commentary (Rome (1936), 341,15-346,15) that he takes Ptolemy to be saying only that only if the stars move in a circle can they maintain a constant distance from the earth. Theon realises that this consideration by itself would not rule out that the cosmos is a right cylinder or cone with the earth on its axis. He then invokes the oblique motion of the planets and shows with simple arguments that if the cosmos were a right cylinder or cone with the earth on its axis then a planet moving on the surface of the cosmos would not maintain a constant distance from the earth. He then claims that the same thing can be proved if the cosmos is assumed to have any other non-spherical shape. It appears that Simplicius is picking up on the contents of this passage in Theon.

239. Simplicius quotes Heiberg (1898-), vol. 1.1, 13,11-12. The *ta* in Heiberg is presumably a misprint for the *tas* printed by Karsten. Pedersen (1974), p. 36, suggests that the reference is to spherical sundials.

240. Heiberg gives *Timaeus* 33B as reference, but there is no very exact correspondence between what Ptolemy says and anything in Plato.

241. Heiberg (1898-), vol. 1.1, 13,22-14,4. The inserted words 'has homoiomerous surfaces' are taken by Heiberg from the text of Ptolemy.

242. Moraux prints *aph' hautou eph' hauto*. At 412,11 and 413,10 (here Karsten prints *apo tauto*), and 14 Simplicius has *apo tou autou epi to auto*. At 413,16 Simplicius has *apo tautou epi to auto* (Karsten prints *apo tou autou*). The words occur again at 414,5, where Heiberg prints *aph' heautou eph' heauto* with A and F, and Karsten prints *apo tautou epi tauton*.

243. The words 'proceeding from the same power' introduce a refinement into Aristotle's dark words 'the fastest motion is the least'.

244. The words 'containing equal plane areas' introduce another refinement into what Aristotle says. There is no least line which goes from a given point and back to it, but if the area enclosed by the line is fixed, a line through the point and containing a circle of that area will be less than any other kind of line which goes from that point and back to it. Cf. the discussion starting at 412,5.

245. cf. *Phys.* 4.14, 223b12-23. But notice that the sense in which the heavenly motion is the measure of motions has no clear relation to the sense in which a measure is a least quantity.

246. Simplicius mentions a Greek measure, the *khous*.

247. Simplicius mentions another Greek measure, the *kuathos*.

248. Simplicius raises the question whether Aristotle could have known that of figures with the same area the circle is smallest.

249. On Zenodorus see *DSB*, vol. 14, pp. 603-5. An epitome of his treatise on isoperimetric figures can be found in Theon of Alexandria's commentary on Ptolemy's *Almagest* (Rome (1936), 354,19-379,15); there are closely related texts in Pappus' *Collection* (Hultsch (1876-1878), 308,2-334,21) and in an anonymous introduction to the *Almagest* (Hultsch (1876-1878), 1138-65). Theon and Pappus refer to Archimedes several times, but it is not known what exactly Archimedes accomplished in this area; see Müller (1953), 44-5.

250. Simplicius describes in an approximate way a simple inference. The theorems he cites say that if a circle or sphere F_1 has a perimeter or surface equal to that of F_2, which is not a circle or sphere, then the area or volume of F_1 is greater than that of F_2. The result he wants for Aristotle is that if a circle or sphere F_1 has the same area or volume as F_2, which is not a circle or sphere, then the perimeter or surface of F_1 is less than that of F_2. For suppose the circle or sphere F_1 and the figure F_2, which is not a circle or square, have an equal area or volume, and that the perimeter or surface of F_1 is not less than that of F_2. Let F'_1 be a circle or sphere with a perimeter or surface equal to F_2 and hence no greater than that of F_1. Then, by the theorems of Archimedes/Zenodorus cited, F'_1 has a greater area or volume than F_2. But F_1 has an area or volume no smaller than F'_1 and hence greater than F_2, but their areas or volumes are assumed to be equal.

251. The meaning of this assertion in the case of solids is not clear. In the case of plane figures, we might consider a square and a circle of equal area and imagine a point running around each under the impetus of an equal 'force'. By the theorem just cited a point will move around the circle in less time than one will move around the square. But the only sense Simplicius has given to saying that the point on the circle is faster is that it makes a circuit around an equal area in less time, but one would expect the relevant magnitude to be the perimeter, not the area. Ultimately the problem would seem to rest with Aristotle's assertion that the fastest motion is along the shortest line.

252. The theorems which Simplicius and apparently Alexander thought Aristotle was using do not depend on the quadrature of the circle in the sense of the construction of a square equal in area to a circle, only on the ability to compare sizes of given figures and lines. Simplicius comes close to making this point at 414,5-11, but the material immediately preceding that is quite confused.

253. See the note on 287a27 in the lemma.

254. Apparently a citation of Euclid, *El.* 1, def. 14.

255. See the note on 287a27 in the lemma.

256. See the note on 287a27 in the lemma.

257. i.e. at 287a11-9; see 409,32-410,15.

258. Translating the *to tên* of F and Karsten rather than the *tên tou* of A printed by Heiberg. Comparison with 412,33 suggests *to tên tou*.

259. *ho stoikheiôtês*, perhaps Euclid, but no proposition in the *Elements* corresponds exactly to the one invoked by Simplicius here. The closest is the more general *El.* 2.14: 'to construct a square equal to a given rectilinear <figure>'.

260. This paragraph is embarrassing. It is true that there is a formal difference between the problem of constructing a straight line or lines equal to the circumference of a given circle and that of constructing a square equal to a circle, but the two problems are mathematically equivalent, since either gives a way of representing pi as a ratio of straight lines. And, of course, 3 1/7 is only an approximation of pi and does not solve either of the problems. See also 549,11-12 in the commentary on Chapter 14 with the note there.

261. See the note on 287a27 in the lemma.

262. Simplicius refers to *Timaeus* 33B1-7, which he has just quoted at 408,31-409,3.

263. Translating Guthrie's punctuation of this text, which seems to represent Simplicius' understanding.

264. In Chapter 14.

265. *hôs pros aisthêsin*. This may be a reference specifically to the arguments at 297b23-298a15, but it is hard to see why Simplicius would refer only to the arguments in that passage. The words *pros aisthêsin* are common in Ptolemy's *Almagest*, as in the title of 1.4: *hoti kai hê gê sphairoeidês estin pros aisthêsin hôs kath' hola merê* (Heiberg (1898-) vol. 1.1, 14,17-18). In any case, as 417,24-6 indicates, the words are probably here because Simplicius knows the earth is not a perfect sphere.

266. Here and in the immediate sequel Simplicius takes up suggestions of Alexander, but the suggestion about the outer surface of air seems unacceptable since on its own the air would move up, not settle in hollows in water. Something like constraint from what is above seems required to force it down.

267. The text says 'inner surface of the moon'. If that is what Simplicius wrote, it was presumably an oversight, or perhaps he is identifying the lunar sphere with the moon.

268. *koiloteron*. As the next sentence and Simplicius' presentation of the argument make clear one point inside a circle is more 'hollow' than another if it is closer to the centre of the circle.

269. In the lemma Heiberg prints *hê BEG*. Our texts of Aristotle read *BEG*, and the *hê* is missing in F and Karsten.

270. i.e. in the last sentence of the lemma. Simplicius is justifiably uncertain what Aristotle's 'central point' is the central point of. See below 417,4-6.

271. Heiberg indicates that only A has the figure, but in it the letters B and A are exchanged and the line DE is omitted.

272. *Neilôiôn kibôriôn*. Simplicius refers to a kind of lotus (*Nelumbo nucifera*) no longer extant in Egypt, but famed in antiquity. The plant is described by Theophrastus (*Historia Plantarum* 4.8.7, on which see the note in Amigues (1989)).

273. The lemma here has *stoikheiôn*. Moraux prints *sustoikhôn*, and 418,3-4 shows that this was Simplicius' reading.

274. cf. 414,23 with the note.

275. In his apparatus Heiberg suggests reading *mêde* for the *mêden* of the MSS. I prefer to keep the *mêden* and drop the *ti* in line 27.

276. Simplicius has *etornôsato* where our texts of Plato have *etorneusato*.

277. *Timaeus* 33B4-5.

278. *Cael.* 1.4, 271a19-22.

279. For Simplicius' understanding of these words see 418,27-31.

280. See 366,9-10 in the outline of Book 2 at the beginning of the commentary.

281. Chapter 2, 286b2-4.

282. In Chapter 2.

283. cf. 394,28-395,3.

284. *hoi eis toupisô anabainontes*. Is the reference to the mounting of a horse starting with one's back to it?

285. Make two right turns?

286. Omitting the *tôi* with K and Karsten rather than including it with Heiberg, following A and a correction from *to* in F.

287. There is a misprint of *ouranon* as *ouanon* in the Heiberg text.

288. See Chapter 2, 285b14-27, with Simplicius' comments at 391,23-392,5 and 394,28-35.

289. Heiberg's period should be replaced with a question mark, as in Karsten.

290. In Chapter 2, 285b28-33.

291. i.e. the necessity that there be coming to be; see Chapter 3, especially 286b2-4. Simplicius is here raising what seems to be a serious difficulty. Aristotle cannot both define forward in terms of the way the heaven moves and say it moves forward because it is best for it to do so. In the next paragraph Simplicius solves the difficulty by saying that Aristotle does not, in fact, use the motions of the heaven to define what is front. In the following paragraph he attempts to give an independent explanation of why the present heavenly motions are best.

292. Reading the *dêmiourgei* of F and Karsten rather than the *dêmiourgousi* printed by Heiberg.

293. The *tetagmena* of F and Karsten seems preferable to the *tetagmenoi* printed by Heiberg.

294. Here is the accepted sequence in which the sun passes through the signs of the zodiac and its correlation with the onset of the seasons:

Aries	Equinox	Spring1
Taurus		Spring2
Gemini		Spring3
Cancer	Solstice	Summer1
Leo		Summer2
Virgo		Summer3
Libra	Equinox	Autumn1
Scorpio		Autumn2
Sagittarius		Autumn3
Capricorn	Solstice	Winter1
Aquarius		Winter2
Pisces		Winter3

Simplicius imagines the sun moving from east to west through the zodiac, producing a backwards year in which the order in which the sun passes through the signs of the zodiac is reversed, producing the following order:

Aries	Equinox	Winter3
Pisces		Winter2
Aquarius		Winter1
Capricorn	Solstice	Autumn3
Sagittarius		Autumn2
Scorpio		Autumn1
Libra	Equinox	Summer3
Virgo		Summer2
Leo		Summer1
Cancer	Solstice	Spring3
Gemini		Spring2
Taurus		Spring1

Simplicius correctly says that in this circumstance after being in Aries the sun would move through Pisces and Aquarius rather than Taurus and Gemini and that it would be summer after the sun is in Libra. But it seems that it should be winter rather than autumn after the equinox in Aries and mid-winter rather than the beginning of autumn when the sun is in Pisces.

295. cf. 394,28-395,3.

296. See 366,10-11 in the outline of Book 2 at the beginning of the commentary.

297. See Chapter 4, 287a23-4.

298. This is rotation on their axes; see the commentary on Chapters 8 and 9.

299. *anelittousôn*. On the 'counteractive' spheres introduced into astronomy by Eudoxus, see the commentary on Chapter 12 at 490,19-507,8.

300. *tois peri Eudoxon and Kallippon*, which could be a way of referring to Eudoxus and Callippus or their followers or both. Lasserre (1966) prints lines 13 to 17 as fr. 122 of Eudoxus.

301. See *Metaph*. 12.8.

302. cf. 425,17-26 and 427,4-14.

303. Simplicius' substitution of 'or' for Aristotle's 'and' makes the claim more plausible; see 425,4-16.

304. Aristotle's division is among naturally moving things, unnaturally moving ones, and 'projectiles', but Simplicius is unable to provide a precise rendering of Aristotle's very sketchy remarks.

305. Simplicius takes projectiles to be living things (see 423,20-36), so the division is among simple bodies moving naturally, simple bodies moving unnaturally, and (composite) living things moving as they ordinarily do.

306. A rather implausible suggestion, but Aristotle's claim is very obscure; see e.g. Drabkin (1935) and Federspiel (1992).

307. Reading *autois* instead of the *autôi* printed by Heiberg.

308. Alexander considers the possibility that projectiles are things thrown in a horizontal direction, that course of motion being 'in the middle' between up and down. He takes back the suggestion at 424,32-4. Perhaps then he ended up holding the same view of projectiles as Simplicius.

309. In order to agree with the word 'motion' (*phora*).

310. e.g. a stone thrown upward.

311. e.g. a stone dropped from a high place.

312. cf. 422,23-8.

313. Heiberg prints *Eti, epei* with F (and Karsten) and says that *Eti* is missing in A. Moraux prints *Eti d' epei*.

314. Heiberg's lemma omits a *to* which is in Aristotle and Karsten, and is included in Simplicius' citation at 425,32.

315. See *Phys*. 7.1.

316. That everything which moves is moved by something.

317. In *Cael*. 1.3.

318. *ek tou mallon*. Simplicius here mentions one of the four argument forms referred to by Aristotle as *ek tou mallon kai hêtton* and described at *Top*. 3.10, 114b37-115a14. The particular form is described at 115a6-8: 'Another <form> is when one thing is said of two: if the thing does not belong to the one to which it is more likely to belong, it doesn't belong to the one to which it is less likely to belong, and if it does belong to the latter, it belongs to the former.'

319. Bracketing *to kinoun* in line 16.

320. See *Phys*. 8.5-6.

321. Simplicius refers to Aristotle's use of the optative, in the translation 'would not move'.

322. cf. 422,8-28 with the note on 422,28.

323. i.e. in the previous lemma.

324. That is, in this lemma and the next.

325. Translating the *autou* of F, not printed by Heiberg. Karsten prints *hautou*.

326. Reading *gê* and *pur* with F and Karsten for the two occurrences of *têi* in A, which are printed by Heiberg; the point is that the elements get mixed together throughout the body.

327. i.e. of their getting sick and decaying.

328. Substituting *kineitai* for *diatithetai*. If Simplicius wrote 'therefore, the divine body is not constituted non-uniformly', he wasn't paying attention.

329. Translating Karsten's *kath' holotêta* rather than Heiberg's *homalotêtos*, which he derives from b. F has *holotêta*.

330. Simplicius is worried about why Aristotle mentions acceleration, when only deceleration would seem to be a consequence of loss of power. Following Alexander, he suggests that acceleration and deceleration are conceptually interdependent, so that what decelerates is also capable of accelerating and vice versa.

331. cf. 288a19-22 in the first lemma for the chapter.

332. I have followed Heiberg in ending the quotation of Alexander with the preceding sentence. Presumably, then, the 'he' in question is Alexander, whom Simplicius is praising. It seems to me possible that 'quotation' of Alexander continues down to the next lemma, and that the 'he' is Aristotle, whom Alexander is praising. The fact that Simplicius writes 'Alexander says' in 429,17 does not rule out that what immediately precedes is a quotation of Alexander.

333. The reader is advised that this passage is for Simplicius only part of a full argument (extending to 289a8) in which three alternatives are refuted, the first being handled in this passage. See 429,22-8, 431,11-22, and 434,21-9.

334. See the note on 466,32.

335. i.e. infinite.

336. The difficulties in the lemma under consideration seem to centre on its last sentence, which Alexander has just quoted. As will be seen, some people wanted to interpret it as a conclusion (obviously an absurd one) reached by Aristotle according to which a deceleration of the heaven would have to be infinite because there could be nothing to restore the power lost by the prime mover which caused the deceleration. Alexander's solution, which Simplicius accepts, is to say that at this point Aristotle is now refuting the first alternative, according to which the heaven accelerates for an infinite time and decelerates for an infinite time, but has failed to mention the infinite acceleration. However, Alexander also accepts the ascription to Aristotle of the argument that a deceleration of the heaven would have to last infinitely long; so does Simplicius.

337. See 429,22-7.

338. In Aeolia; modern name: Nimrud-Kalessi. On this Alexander see Moraux (1984), pp. 222-5, whose discussion has been very helpful to me. Presumably Alexander took the point of the last sentence to be that an infinite deceleration would rule out an acceleration. He then apparently inferred that there would have to be an infinite acceleration before the infinite deceleration.

339. On Herminus see Moraux (1984), pp. 361-97, and for Aspasius, pp. 226-93.

340. 288b27-8; see 432,15 ff.

341. 289a4-5; see 434,19 ff.

342. That is, the hypothesis of this part of the division is really that there is both an infinite deceleration and an infinite acceleration, but Aristotle also argues that if there is an infinite acceleration and a deceleration, the deceleration will also be infinite.

343. This argument would seem to apply only to rectilinear motion. Simplicius' attempt to extend it to circular motion in the next paragraph does not appear to be successful. But it is not clear how Aristotle would apply his generalisation about motion to the motion of the heaven.

344. The standard punctuation of this sentence would have no stop here, but an open parenthesis closed after the word 'similar'. On this reading the final clause, 'if it accelerates ...' is a recognition that a continuously decreasing non-uniform acceleration is compatible with the speed having a limit. I have punctuated according to what I take to be Simplicius' understanding, which does not involve a recognition of this point; see 433,31-434,18.

345. Moraux prints *tôi autôi te* here. In a citation at 434,14 Heiberg prints *tôi autôi*, Karsten *tautôi t'*.

346. Translating the *auto* of A and Karsten instead of the *autôi* of C which Heiberg prints, noting that F corrects *auto* to *autôi*.

347. Simplicius' remarks in this paragraph are less fine-grained than what he says at 428,32-429,16.

348. Presumably Simplicius has in mind that a thing increases its speed by covering ten finger's breadths more in the same time, then nine more than that, and so on. On the use of the 'finger' as a unit of measurement in astronomy see Neugebauer (1975), pp. 658-9.

349. With this paragraph see the note on 289a2 in the lemma.

350. The *hêmin* which is included in the MSS of Aristotle is not printed by Heiberg, following A. It is included in F and Karsten.

351. Heiberg prints *asterôn*, where Moraux and Karsten have *astrôn*.

352. See 366,11-12 in the outline of Book 2 at the beginning of the commentary, and note that the seventh topic covers all of Chapters 7-11.

353. In Chapter 12.

354. Simplicius summarises *Cael.* 1.2.

355. Simplicius recalls the view he ascribes to Plato, which he has set out in discussing *Cael.* 1.2-3. See the note on 375,2.

356. Heiberg prints *ek tou puros* where Rivaud (*Oeuv. Comp.*, vol. 10) has *ek puros*. The *tou* is in A, but is omitted by F and Karsten. See also 454,29-31.

357. *Timaeus* 40A2-4. For the continuation of this passage see below 454,29-455,8.

358. The remainder of the discussion of this lemma concerns a difficulty raised by Alexander: (a) What accounts for the apparent difference between a star and the sphere on which it is carried when they are both made of the same substance? Alexander suggests that the difference must be due to qualitative differences or oppositions, and then argues, in what might be called a conceptual way, that (b) the presence of such qualitative differences does not mean that heaven is subject to change. Although Simplicius ultimately agrees with Alexander on (b), he does not like Alexander's explanation of it and offers his own more metaphysical account, starting at 436,27. But, having decided that qualitative differentiation is compatible with unchangingness, Simplicius insists that question (a) remains open since it is not clear why there should be differentiation if heaven is composed of one uniform substance. His brief discussion at 438,13-26 is not earth-shattering or mind-boggling: sublunary things come to be and cease to be because they have a substratum which moves in a straight line, and they have all sorts of different qualities; heavenly things do not come to be or cease to be because their subtratum moves in a circle, but nothing prevents their having different qualities. The next section of the commentary should be compared with 372,24-373,24.

359. Alexander quotes *Cael.* 2.1, 284a14.

360. Reading Bessarion's correction *to leukon kai psukhron*, which is also printed by Karsten, rather than the *to leukon to psukhron* printed by Heiberg, following A and F.

361. On these constellations see Kidd (1997), pp. 265-6 and 204-5.

362. Putting a period after *Alexandrou*, inserting the *hoti* of F and Karsten (omitted by A and not printed by Heiberg), and eliminating Heiberg's colons in lines 27 and 28.

363. For a brief discussion of the complicated sense this term takes on in Neoplatonism see Wallis (1995), pp. 67-70.

364. Presumably what Simplicius has in mind is what we might call the interrelatedness of opposite concepts; anyone who understands a concept like hot also understands its opposite, cold.

365. i.e. the spheres.

366. Simplicius virtually quotes *Cael.* 1.3, 270a14-17. 'Our first discussions' would seem to be a reference to *Phys.* 1.7-9, as Simplicius goes on to suggest here. See also Simplicius' commentary on *Cael.* 1.3, 270a12-22.

367. The verb is in the present tense, but Simplicius is referring to what Aristotle argued in *Cael.* 1.3.

368. Moraux has *auta* (missiles) here, but Simplicius cites as *hautai* (lead balls) at 439,21. For other expressions of this incredible notion of missiles ignited (or at least melted) by motion see Guthrie ad loc.

369. *tas molibdidas tas ton sidêron en tôi xulôi sunekhousas.* I do not know what to make of this description.

370. The *hupekkauma* is a highly flammable vapour at the upper regions of the sublunary world invoked in the *Meteorology* to explain various phenomena, notably for present purposes the heat and light we associate with heaven. The ancient commentators seem to have been agreed that *Meteor.* 1.3-4 makes clear that the *hupekkauma* is somehow related to air and fire, but the description is not terribly precise. In the present passage Aristotle seems to assign to air the role he assigns to the *hupekkauma* in the *Meteorology*, and this leads to Simplicius' difficulties. In the present sentence Simplicius offers his 'paraphrase' of what Aristotle says; according to it, the motion of the heaven sets the *hupekkauma* on fire and that heats the air beneath it.

371. It appears from 439,27-8 that Alexander actually said that Aristotle is now calling air the sum of the *hupekkauma* and the upper air.

372. For Alexander (cf. *Quaest.* (Bruns (1892)) 2.17, *Mant.* (Bruns (1887)) 148,20-3, and *in Meteor.* (*CAG*, vol. 3.2) 17,26-8) the *hupekkauma* which Aristotle describes in *Meteor.* 1.3-4 is the element fire, the fire around us being always mixed with something burning. At 17,18 of *in Meteor.* Alexander, perhaps thinking of the present passage, says that it is also Aristotle's custom to call the *hupekkauma* air. Alexander (cf. *in Meteor.* 16,32-17,1) thinks the doctrine of *Meteor.* 1.3, 341a12-28 is the same as that expounded here with the word 'air' substituted for *hupekkauma*; so too Philoponus, *in Meteor. 1* (*CAG*, vol. 14.1) 39,29-30. Simplicius resists the assimilation, suggesting that Aristotle here adds to the account of the *Meteorology*, in which he describes how the *hupekkauma* is heated from above, the point that the air beneath the *hupekkauma* is heated by it.

373. cf. 373,1 ff.

374. For this description of what is moved by the heaven cf. Aristotle, *Meteor.* 1.7, 344a11-13.

375. For a critical treatment of Aristotle on this point see Philoponus, *in Meteor. 1* (*CAG*, vol. 14.1) 41,15-43,33; he argues that the heat should be associated with light, not motion. Olympiodorus discusses the same question at *in Meteor.* (*CAG*, vol. 12.2) 32,9-34,5.

376. cf. *Meteor.* 1.3, 341a23-7.

377. i.e. the *hupekkauma* and the upper air.

378. i.e. the sun.

379. i.e. the sun. Heiberg omits the words 'of the body' (*tou sômatos*), which are found in F and Karsten. Alexander is offering an explanation of why the sublunary region near the sun is warmer. His answer is that the sun is a large,

dense body not far from the sublunary region. Aristotle invokes the closeness of the sun in his explanation at *Meteor.* 1.3, 341a12-28. I have not found any discussion of this exact topic in Alexander's commentary on the *Meteorology*, but at 19,13-19 he explains the greater coolness of the shade in terms of the intercession of the object casting the shadow.

380. See 373,1 ff. Note that in our present passage Simplicius rejects Alexander's account. He does not want there to be pushing and resistance in heaven, and he does not think an effect can be transmitted without pushing and pulling.

381. I have translated Heiberg's text (*sômatikas*), but I suspect Simplicius wrote 'incorporeal' (*asômatikas*); cf. Olympiodorus, *in Meteor.* (*CAG*, vol. 12) 32,11-13.

382. The equal angles are important to Simplicius because the higher the sun is the more acute the angles will be and so the more the air will be compressed. A similar line is taken by Olympiodorus, *in Meteor.* (*CAG*, vol. 12) 32,20-33,7.

383. A comma before *aei* in line 14 would help.

384. i.e. obtuse.

385. Simplicius paraphrases *DA* 2.7, 418b11-13. Philoponus has an interesting discussion of the production of light and heat in his commentary on this passage (*in DA* (*CAG*, vol. 15) 324,25-329,29). For him light involves incorporeal rays which activate both the transparent and the heat in air. There is no question of motion and rubbing. In his discussion of vision in *Mant.* (Bruns (1887)) 141,30 ff. Alexander takes an analogous position for light with no mention of its being produced by rubbing.

386. See *Meteor.* 1.3, 341a12-17.

387. But see 443,21-4 for Simplicius' strained interpretation of what Aristotle means.

388. This is not something Aristotle says, as Simplicius points out at 443,18-21.

389. In Chapter 2.

390. It is tempting to think that some words have dropped out here.

391. Simplicius objects to this odd characterisation of circular motion as a tangible quality at 444,4-6.

392. *GC* 2.2, 329b18.

393. *GC* 2.2, 329b19-21.

394. Presumably density.

395. cf. 441,21-6.

396. *Timaeus* 31B4-6, but Plato speaks of earth and fire, not of what is earthy (*gêinos*) and fiery (*purinos*).

397. cf. 437,9-29.

398. Simplicius picks up again on 284a14.

399. *kathistamena palin*. Simplicius prefers the more technical *apokathistasthai*; see e.g. 445,14.

400. On the text here see the note on 447,27. Here and elsewhere Aristotle only considers linear, not angular, velocity. For him the equator moves at a greater speed than the tropics.

401. Translating the *apo tukhês* of F; see 446,31-447,6 with the note on 447,5. Karsten and Moraux have *to apo tukhês*, giving the sense 'what is a matter of chance does not hold everywhere and in all cases'. Heiberg prints *tês tukhês*, presumably because A has the non-sensical *tês psukhês*; I do not know how Heiberg would construe *tês tukhês*.

402. This is the daily east-west revolution of the fixed stars.

403. The example of changes given here are a result of the sun's west-east motion. Simplicius wants to avoid the idea that what Aristotle is talking about applies to the special characteristics of planetary motion; see 447,20-448,2. For him, I think correctly, the real issue concerns the relative motion of a star and the sphere on which it rests, and can be discussed entirely in terms of the fixed stars and their sphere. Given a stationary earth, the question is whether they both move or just one does (and, if one, which one).

404. For an account of these people and their astronomical views, see e.g. Heath (1913), Part I, Chapter 18, and Part II, or, for Heraclides, Gottschalk (1980), Chapters 1 and 4. Heath's conclusion, widely accepted, is that although, as Simplicius here suggests, both men held that the earth rotated on its axis, Aristarchus made the earth and the other planets revolve around a central sun, whereas Heraclides retained a basically geocentric system but made Venus and Mercury revolve around the sun.

405. i.e. its daily motion through the zodiac.

406. In Chapter 14.

407. It is not clear to me what possible 'observation' Simplicius has in mind, since the circles or spheres are not observable at all and the stars are only visible at night. But, if we imagine all the other stars on the circle moving at the same speed as the circle, then the star would be seen departing from its 'place' on the circle. Perhaps something has dropped out of the text here.

408. In this chapter at 290a1-2. Simplicius imagines that the stars move around on their own naturally. He thinks this means that greater stars will be moving in greater circles. So, if the stars are imagined exchanging their circles, they would either have to move with different angular velocities or their motion would no longer be natural.

409. Heiberg prints *de ouk* with A instead of the *de kai ouk* of Moraux and F (Karsten has *de kouk*).

410. Heiberg follows F in printing *oude to pantakhou kai pasin huparkhon apo tukhês*, words of which there are no trace in A. Karsten prints *to apo tukhês*. The first of Simplicius' two syllogisms suggests he read *to apo tukhês*, the second that he read *apo tukhês*; see also the last note on the lemma.

411. Heiberg starts a new paragraph here, implying that the next paragraph is not Alexander. Certainly there is no way to tell with certainty, but the phrase 'necessity of the circles' sounds to me more like Alexander than Simplicius.

412. Reading the *auton* of F and Karsten rather than the *autôn* printed by Heiberg.

413. Aristotle's use of 'circle' in the lemma is, indeed, nebulous. But Alexander's introduction of the planetary spheres here invites confusion, as what follows shows. And Alexander persists in bringing in the planets in connection with what Aristotle says in this chapter. See 450,6-451,27.

414. Simplicius' citation has a *kai* which is not in Moraux's text and omits a *gar* which is in it.

415. It is difficult to know what to make of this remark. One might assume that Alexander was making some assumption about the distances of the planets from the earth so as to make their linear velocities proportional to their distances or (what seems more likely to me) that he is talking about angular velocities and means to say that they are inversely proportional to the distances. In any case Simplicius rebuffs Alexander.

416. See 448,27-449,2, which casts serious doubt on Simplicius' representation of Alexander here.

417. Simplicius wants to retain the 'psychic autonomy' of the stars; see 454,23-456,27.

418. For Simplicius' explanation of this term see 448,28.

419. Heiberg's lemma (but not Karsten's) omits the last words *kata ta megethê tôn kuklôn*, but they are included in the citation at 448,28-9.

420. For the reasoning see 446,15-31 with the note on 446,22.

421. The force of this remark is not clear to me, and I am tempted to think Simplicius meant to say that a smaller star would not be moving on a smaller circle, meaning that if the sizes of the circles are responsible for the speed of the motions there is no necessary correlation between size of star and speed.

422. I have followed Karsten in inserting <*tôn kuklôn*>.

423. Heiberg prints *Epeidê* with A; F and Karsten have *Epei*, which is printed by Moraux.

424. *tôn ek tou kentrou. hai ek tou kentrou* <*eutheiai*> are radii, but at 452,2-6 Simplicius indicates that Alexander espoused taking the words to refer to planes drawn through the central point. Perhaps the issue can be explained in the following way. The simplest way to understand what Aristotle says is in terms of concentric circles in the same plane revolving at equal angular velocities, say with radii AB and ABC. Aristotle's point is that C moves 'faster' than B because it moves through longer arcs in equal times. The trouble is that this picture doesn't apply to the fixed stars, which move on parallel circles in a single sphere. One way out might seem to be to say that the reference was to the west-east revolution of the more or less concentric circles of the planets, but those motions do not obey the proportionality rule, as Alexander is quoted as saying just before this lemma. The other way is to try to read central 'point' by reference to the axis of the sphere of the fixed stars. Note that this problem also applies to the words 'fastened about the same central point'.

425. Aristotle is perhaps referring to 287a2-11 and 287a30-b14, but, as Alexander points out, the universe is not continuous in the strict sense. See 451,7-13 and 466,29 with the note.

426. Although the desired sense is clear enough, something seems to have gone wrong here. I have read the two *ta* in line 26 as *to* with F and Karsten against A and C, and bracketed the *to* in the next line.

427. I have not found a satisfactory way to render this grammatical point in English. There are three occurrences of *te* in the passage. I have represented the correspondence pointed out by Simplicius by inserting 'first' and 'second' in the lemma. The correspondence he wishes to rule out is ruled out by my use of 'both' and 'and' in the translation of the last sentence of the lemma.

428. Simplicius expresses perplexity with this remark of Alexander's at 450,30-451,1. Alexander's suggestion is perhaps that the speeds of planetary circles are not determined by the size of their circumferences, which is a purely mathematical fact, but by the physical truth that greater bodies have faster natural motions. But in the next citation he indicates that Aristotle's argument doesn't really apply to the planets.

429. i.e. their west-east motion.

430. Reading *autêi têi* with F and Karsten rather than the *têi* printed by Heiberg. Simplicius assigns to Alexander the view which he himself accepts: that the circles Aristotle is talking about are latitudes on the sphere of the fixed stars. But he does not tell us how Alexander gets this reading out of the text. His own proposal involves a very loose reading: Aristotle cites a plane geometric theorem about the arcs of concentric circles cut off by radii, but he really has in mind the arcs of latitudes on a sphere cut off by longitudes.

431. Heiberg prints *sunekhes* with A rather than the *sunekhes on* of F and Karsten and our MSS of Aristotle.

432. Alexander points out that the heaven is not continuous in the strict sense (see 466,29 with the note), and that disparate motions of the planetary spheres would not cause the cosmos to be 'torn apart'. He goes on to suggest that Aristotle may be speaking about the continuity of each sphere with itself and saying that a sphere containing a planet would be torn apart if the planet moved on its own while its sphere remained fixed.

433. Reading *kinoumenoi* with Karsten. Heiberg prints *sunkinoumenoi* with A. F has *hoi kinoumenoi*. Cf. 451,18.

434. At 289b27-30. According to Simplicius the argument now concerns the relative velocities of the circles on which the fixed stars have been shown to be carried.

435. cf. 450,17-24.

436. Translating Heiberg's conjecture *têi sphairai*. He reports that all of A, F, and Karsten have *tês sphairas* (if the circles of the sphere have not been engraved). It is unfortunate that the MSS show considerable variation and uncertainty in this paragraph in which Simplicius discusses the ontological status of the circles described by the fixed stars.

437. I have translated the text of F and Karsten: *kai houtoi eisin hous legei ho Aristotelês analogon ekhontas ta takhê tois megethêsin*. For *eisin* ... *ekhontas* A has *ei logon ekhei tauta*, for which Heiberg conjectures *hoi logon ekhontes* .... A reader has pointed out that Heiberg would have done better to omit *hoi* and place a comma rather than a period before *kai*.

438. Translating the citation of this sentence by Simplicius at 477,8-12 in the commentary on Chapter 11: *eti de, epei sphairoeidê ta astra, kathaper hoi te alloi phasi kai hêmin homologoumenon eiper ex ekeinou ge tou sômatos gennôsin* .... Our texts of Aristotle have *eipein* where Simplicius has *eiper*, giving a text which Leggatt translates as 'since the stars are spherical, as others maintain and as we may declare with consistency, given that we generate them from that body'.

439. *kulisis* and *dinêsis*. Simplicius explains these terms at 452,18-19. Roughly speaking, a ball rolls on the ground but a globe rotates on its axis. Simplicius' position is that a spherical star rotates on its own while being carried around in a circle. The effect of these two motions is of course no different from rolling on a circular path, but it is important for him to insist that the two are not the same thing as rolling. Cf. 457,2-5.

440. I do not know what phenomenon Aristotle is referring to here. Leggatt ad loc. refers to Xenophanes' view (DK21A41a) that the sun really moves in a straight line, but appears to move in a circle because of its distance from us. But it seems that Xenophanes has in mind the apparent motion of the sun around the earth, not its rotation.

441. *endedemenous*, i.e., fixed, as Simplicius explains at 452,12-14.

442. On this word see 453,34-454,4 with the note.

443. *tous menontous*, i.e. the fixed stars, as Simplicius explains at 452,18-19.

444. In Chapter 11.

445. This condition is crucial for Simplicius. Rotation would not attach to the stars if the spheres were at rest since rotation alone would not account for their change of position. This, however, is quite compatible with their rotating on their own while being carried around by their sphere.

446. But Simplicius disagrees.

447. Translating the *parapleontes* of F and Karsten rather than the *parapempontes* of A, printed by Heiberg. A reader has suggested that one might substitute the passive *parapempomenoi* ('coast past') or translate the active as 'let the shores go by them'.

448. i.e. near the horizon.

449. The word which I have translated 'vision' (*opsis*) can also be translated 'ray'. However it is translated, Aristotle does appear to be relying on a conception of vision according to which it involves the emission of something from the eye, whereas the theory standardly assigned to him on the basis of *DA* 2.7 involves the emission of a 'form' from the object, and in *De Sensu* he explicitly denies that there is an emission from the eye; see 454,6-14 with the note.

450. The *êkousen* of F and Karsten seems more plausible than the *êkouen* of A printed by Heiberg.

451. The word 'our' does not occur in our text of Aristotle.

452. *Sens.* 2, 438a25-b2.

453. I do not know on what ground Simplicius makes this assertion. Pachymeres (Failler (1999), 7.16), who knows the theory that the planets do not twinkle because of their proximity to the earth, describes a (miraculous) twinkling of the sun at midday.

454. *ho stilbôn*: the glitterer or twinkler. The name goes back to at least 262 BC (Ptolemy *Almagest* 9.7, 264), but why the Greeks thought Mercury twinkled is not clear. I know no other classical source for the twinkling of Venus, but cf. Eustathius (van der Valk (1987)), 622,10, who says (probably mistakenly) that Venus is also called *stilbôn*.

455. On the colours of the planets see PW vol. 20, cols 2105-6.

456. Heiberg prints *apeirgasato*, following A and F; Karsten has *apeirgazeto*; Rivaud (*Oeuv. Comp.*, vol. 10) prints *apêrgazeto*. See also 436,1-3.

457. Heiberg prints a *kai,* which is omitted by Karsten and is not in Rivaud's text (*Oeuv. Comp.*, vol. 10).

458. I have translated Rivaud's text (*Oeuv. Comp.*, vol. 10): *tên men en tautôi kata tauta, peri tôn autôn aei ta auta heautôi dianooumenôi*. Heiberg prints *tên men en tautôi kata tauto dinoumenôi*, basically following A. The words *kata tauta, peri tôn autôn aei ta auta heautôi* are found in F and Karsten except that F has *tauto* and Karsten *tauth'* for *ta auta*. The *dinoumenôi* is Heiberg's conjecture; Karsten agrees with Rivaud, A has *dianooumenon*, F *dinoumenon*.

459. Simplicius has *hestôs* where Plato has *hestos*.

460. Backward, upward, downward, to the right, to the left.

461. Heiberg prints a *kai* which is not in Plato or Karsten.

462. Rivaud (*Oeuv. Comp.*, vol. 10) and Karsten print 'divine animals' (*zôia theia*); Heiberg, following A and F, omits the *zôia*.

463. Translating Heiberg's *dia tauta en tautôi strephomena*. Rivaud (*Oeuv. Comp.*, vol. 10) prints *aidia kai kata tauta en tautôi strephomena*, as does Karsten (except for two elisions).

464. *Timaeus* 40A2-B8.

465. Simplicius omits the *men* which is in Aristotle here and at 456,16 and 31. In the first case Karsten prints the *men*, but not in the last two.

466. Moraux prints *ton topon*, Heiberg *topon*.

467. Simplicius introduces 290a33 ff. See below 458,8 ff. with the notes.

468. Heiberg prints *katekhei*, where Karsten and the MSS of Aristotle have *katekhoi*. But Heiberg prints *katekhoi* (corrected from *katekhei* by a second hand in E) in a citation at 478,24 in the commentary on Chapter 11, and in a paraphrase below at 459,1 (with A, E, and Karsten; D and F have *katekhei*).

469. See the note on the word 'stars' at 290b9 in the lemma below at 450,9.

470. i.e. in the sphere of the fixed stars.

471. *Cael.* 2.12, 292a20.

472. See the next lemma.

473. Simplicius does not have the *men* of Aristotle. Similarly at 456,31.

474. That is, I take it, to say that the stars do not rotate while standing still is not to say that they don't rotate while changing place, but Aristotle's argument only eliminates rotation while standing still.

475. But one might still hold that the sun rotates in a way which is not visible to us.

476. The passage quoted here does not survive in a Greek text of Ptolemy, but it is preserved in an Arabic translation, rendered into German by Ludwig Nix (Heiberg (1898-), vol. 2, 131,9-15). The Arabic text is printed by Goldstein (1967). I reproduce Nix's translation: 'Es ist also richtiger, dass jedes einzelne Gestirn auch etwas bewege, weil dies die Kraft und das Wirken des Gestirns an seinem ihm eigentümlichen Platze und um seine Mitte ist, nämlich die zusammenhangende drehende Bewegung. Es ist also notwendig, dass der Anfang der Sache vom Gestirn ausgeht, indem es sie durch die Körper ausführt, die sie umgeben.'

477. Heiberg's lemma has *kinointo*, Karsten's *kinoito*, which is what Moraux prints. See also 458,7 with the note.

478. The theory that what we see on the moon is a reflection of the earth is ascribed to Clearchus, a pupil of Aristotle, in sections 3-4 of Plutarch's *Concerning the Face Which Appears in the Circle of the Moon* (Cherniss (1957)). On Iamblichus see Dillon (1987); I have not found a source for what Simplicius says about Iamblichus or the 'others'.

479. Simplicius mentions the one way in which the moon could rotate and always present the same side to the earth, namely if it rotated once for each time it completes a west-east revolution.

480. Heiberg prints *kinoito di' hautôn*, which agrees with our text of Aristotle. A has *kinoito di' autôn*, F and Karsten *kinointo di' hautôn*. Cf. the lemma with the note.

481. Simplicius cites this passage at 478,21-6 in the commentary on Chapter 11, where Heiberg prints *kinoumenois*, Karsten *kinêtikois*, which is in our text of Aristotle.

482. At 455,22 Simplicius cites these words as *ta de alla astra* (here Karsten prints *talla de astra*) and at 460,5 he cites them with *ta de alla ourania sômata, toutesti ta astra*. Moraux thinks that both *astra* and *ourania sômata* are glosses and prints *ta d' alla [astra]*.

483. cf. 460,10 where Simplicius mentions texts which say 'the one will cause motion and the other will be at rest'.

484. i.e. Aristotle is not contrasting heavenly bodies with animals (since they are animals), he is contrasting them with ordinary animals.

485. I propose emending Heiberg's *tôn isoperimetrôn khôriôn elakhistê hê peripherês* ('of isoperimetric areas the round is least', which is false) to *tôn ison khôrion periekhousôn grammôn elakhistê hê peripherês*, which is true and directly relevant to what Simplicius is saying.

486. cf. 411,10-417,17 with the notes.

487. Simplicius cites *Politicus* 270A8 with the words *epi smikrotatou podos aei bainon*. Diès (*Oeuv. Comp.*, vol. 9.1) prints *epi mikrotatou bainon podos*.

488. Simplicius appears to be thinking of animals in particular and not non-spherical things in general.

489. Reading the *kath' hauton* of A rather than the *kat' auton* printed by Heiberg.

490. As Aristotle says at *Cael.* 1.9, 279a11-18.

491. i.e. *Phys.* 4.1-5; see especially 4.4, 211a14-29.

492. See the note on the word 'stars' at 290b9 in the lemma.

493. Moraux (1954) reports that this reading (*kinêsei* instead of *kinêsetai*) is present as a variant in one manuscript (Urbinas 37). Simplicius presumably prefers this text because the other might be taken to imply that only the fixed sphere moves, whereas for him the stars themselves rotate in the same way it does. But in the next paragraph he appears to accept the standard text while qualifying its meaning.

494. It is true *a fortiori*, since the heaven does not progress at all.

495. That is, Aristotle has characterised how things look, not how they are.

496. At 289b13-16.

497. In *Cael.* 1.2, 268b17-269a2.

498. In *Phys.* 4.6-9.

499. On the *hupekkauma* see the note on 439,12. In the present passage Simplicius dismisses the possibility that the spheres might be made of something like the vapour occupying the upper regions of the sublunar world.

500. Simplicius imagines that the moon moves with its 'apparent' motion on its own. It then has the compound motion of its two circles, that is a helix.

501. i.e. sphere.

502. *Almagest* 7.2 (Heiberg (1898-) vol. 2.1, 12,4-16,11). Proclus (*Hypotyposis* (Manitius (1909)), 234,7-23 and *in Tim.* (Diehl (1903-1906)), vol. 3, 124,18-126,5) denied this 'precession of the equinoxes'. On Hipparchus see *DSB*, vol. 15, pp. 207-22. Lasserre (1966) prints this sentence as D15b of Eudoxus.

503. *stereou astrolabou.* On Ammonius see Goulet (1989-), vol. 1, pp. 168-9. The one degree per 100 years is Ptolemy's constant of precession. If Ammonius worked with Ptolemy's figure for the longitude of Arcturus, then the modern determination of its longitude in 500 AD implies that his observation was off by close to 3°; see Neugebauer (1975), p. 1037.

504. This sphere is added to be able to account for precession.

505. This passage was used by Merlan (1968) as part of an argument for the relative chronology of this commentary and Simplicius' commentary on the *Enchiridion* of Epictetus. For a refutation of the argument see Hadot (1978), pp. 40-3.

506. D, E, F, and Karsten all have *homonumôs*, a standard expression in Aristotle and the commentators. Heiberg prints *homonuma*, presumably because A has *kai homonuma*.

507. Simplicius adapts a Homeric description of the sun (*Iliad* (West (1998)) 3.277; *Odyssey* (van Thiel (1991)) 11.109 and 12.323). On the controversy on the question whether Aristotle assigned perception to the heavenly bodies see the discussions of *De Anima* 3.12, 434a30-b8 in [Simplicius] (*CAG*, vol. 11, 320,18-39), Philoponus (*CAG*, vol. 15, 595,33-598,6), and Themistius (*CAG*, vol. 5.3, 123,20-35).

508. *kompsôs kai perittôs.* For Simplicius' explanation of these terms see 465,3-5.

509. At 464,27 in a paraphrase/citation Simplicius has *edokei*. Moraux prints *dokei*.

510. That is, change position; the stars do rotate on their own for Simplicius and Simplicius' Aristotle.

511. In the next lemma.

512. cf. Chapter 8, 289b15-16 and 289b34-296a5.

513. Here and at 464,25 Simplicius speaks of speed and slowness.

514. In the quotation Aristotle uses the word *phônê*, for which, in the words I have not translated (*hêtoi ton psophon*), Simplicius substitutes *psophos*.

Aristotle uses both words, apparently interchangeably, but hereafter Simplicius uses *phônê* only twice where he is quoting or paraphrasing Aristotle (464,28.31). Perhaps the reason is the association of *phônê* with voice. I have translated both words as 'sound'.

515. Moraux, following a suggestion of Bonitz, prints *legein* here, where all the MSS of Aristotle have *luein*. Simplicius offers an 'explanation' of *luein* at 465,18-24.

516. *pros*. Simplicius explains the meaning at 466,2.

517. The point is that sound is produced by impact. Cf. Simplicius' paraphrase at 466,16-17 and *DA* 2.8, e.g. at 419b9-11. For Simplicius the heavenly 'sound' does not involve an impact; see below 468,26-469,20.

518. Aristotle here states two points from the preceding chapter, confirming Simplicius' notion that the present chapter is subsidiary to its predecessor.

519. It seems to me preferable to keep the *auton* of D, E, F, and Karsten omitted by Heiberg, following A.

520. Themistius, *in Cael.* (*CAG*, vol. 5.4) 117,13-17, assigns a similar consideration to Alexander of Aphrodisias. For this and other parallels see Moraux (2001), p. 186.

521. That is, change position.

522. Simplicius has a plural where Aristotle has a singular (counted as a citation by Moraux).

523. Simplicius substitutes *kath' hauta* for Aristotle's *auta*.

524. *Phys.* 4.4, 212a16-20.

525. The kinds of things Aristotle is talking about are not continuous in the strict sense specified at *Phys.* 5.3, 227a10-17.

526. I am not sure what list of hypotheticals Simplicius is referring to here. The phrase 'by means of three' is undoubtedly to be filled out as 'by means of three hypotheticals' (*tropikoi*). On this subject see Frede (1974), pp. 181-3. Simplicius' argument is presumably of the form 'if p then q, if q then r, therefore if not r then not p' (although it might be of the form 'if p then q, if q then r, not r, therefore not p), which suggests that for him the first hypothetical would be 'if p then q, if q then r, therefore if p then r', which is frequently referred to as 'by means of three' (references in Frede (1974), p. 183). The argument which Simplicius gives does not coincide with any of the forms mentioned by Frede, but it is listed as the 59th of some 138 syllogisms by Boethius in *On Hypothetical Syllogisms* (Obertello (1969), 2.11.13-15 (p. 314)). Simplicius refers to the 'second hypothetical mode' in two other passages in the commentary on Book 2. At 429,28 the argument is quite similar to the one here, but at 536,21 in the commentary on Chapter 14 it is simply *modus tollens*, the second Stoic unprovable.

527. For a brief discussion of Alexander's conception of divine providence (*pronoia*) see Sharples (1987), pp. 1216-18.

528. i.e. god, the maker of universe.

529. Moraux ((2001), p. 199, n. 99) assigns the content of this sentence to Alexander rather than Simplicius because Alexander denies that the heavenly bodies perceive, but Simplicius inclines to think they do; see 463,1-12 with the note on 463,12.

530. Reading the *ekei huparkhontos* of D, F, E (as a correction), and Karsten rather than the *enhuparkhontos* of A, printed by Heiberg.

531. Alexander recalls Chapter 4, 287b15-16.

532. Heiberg follows A in omitting the words *tou Aristotelous kai* ('by Aristotle and'). The words are found in E and F, and printed by Karsten.

533. The *metabatikôs* of D, E, F, and Karsten seems preferable to the *metabatika* of A, printed by Heiberg; cf. 470,24, 477,24, 478,10.

534. Simplicius now attempts to explain why the story about Pythagoras hearing the harmonious 'sound' of the spheres is true.

535. I have followed the advice of a reader in translating this long conditional clause and in restoring the *augoeides* of A, D, E, F, and Karsten for the *autoeides* conjectured by Heiberg on the basis of *autoideale* in b. I also adopt the reader's more tentative suggestion that one should substitute genitives for accusatives and read *tou augoeidous autou kai ouraniou okhêmatos*.

536. On the luminous heavenly vehicle and its unified senses see Proclus, *in Tim.* (Diehl (1903-1906)) vol. 2, 81,19-28. There is a summary of ancient ideas about it in Kissling (1922), pp. 318-26.

537. On the importance of ritual perfection see Iamblichus, *Myst.* (Des Places (1996)) 229,12-230,14.

538. I have inserted the *isôs* of D, E, F, and Karsten, omitted by Heiberg, who follows A.

539. In Chapter 11.

540. Heiberg prints *heautôn*, following A, where Aristotle, Karsten, and a second hand in E have *hauton* (D and E have *autôn*).

541. i.e. and not their rotation.

542. Simplicius virtually quotes 290b15-16, where Aristotle says that some people think this.

Bibliography

Adler, Ada (ed.) (1928-1938), *Suidae Lexicon*, 5 parts, Leipzig: Teubner.

Allan, D.J. (1950), 'Mediaeval Versions of Aristotle "De Caelo" and of the Commentary of Simplicius', *Mediaeval and Renaissance Studies* 2, 82-120.

Amigues, Suzanne (ed. and trans.) (1989), *Théophraste, Recherches sur les Plantes*, vol. 2, Paris: Belles Lettres.

Bergk, Theodor (1883), *Fünf Abhandlungen zur Geschichte der Griechischen Philosophie und Astronomie*, Leipzig: Fues's Verlag.

Bodnar, István M. (1997), 'Alexander of Aphrodisias on Celestial Motions', *Phronesis* 42, 190-205.

Bossier, Fernand (1987), 'Traductions Latines et influences du commentaire in *De Caelo* en occident (XIIIe-XIVe s.)', in Hadot (1987), pp. 289-325.

Braunlich, Alice F. (1936), ' "To the right" in Homer and Attic Greek', *American Journal of Philology* 57, 245-60.

Bruns, Ivo (ed.) (1887), *Alexandri Aphrodisiensis praeter commentaria scripta minora, De Anima Liber cum Mantissa* (*Supplementum Aristotelicum* vol. 2, pt 1), Berlin: Georg Reimer.

Bruns, Ivo (ed.) (1892), *Alexandri Aphrodisiensis scripta minora, Quaestiones, De Fato, De Mixtione* (*Supplementum Aristotelicum* vol. 2, pt 2), Berlin: Georg Reimer.

Cherniss, Harold (ed. and trans.) (1957), *Concerning the face which appears in the orb of the moon*, in *Plutarch's Moralia*, vol. 12, Cambridge, MA: Harvard University Press and London: William Heinemann.

Des Places, Edouard (ed. and trans.) (1971), *Oracles Chaldaïques*, Paris: Belles Lettres.

Des Places, Edouard (ed. and trans.) (1996), *Jamblique: Les Mystères d'Égypte*, Paris: Belles Lettres.

Diehl, Ernestus (ed.) (1903-1906), *Procli Diadochi in Platonis Timaeum Commentaria*, 3 vols, Leipzig: Teubner.

Dillon, John (1987), 'Iamblichus of Chalcis (*c.* 240-325 AD)', in Wolfgang Haase (ed.), *Aufstieg und Niedergang der Römischen Welt*, II 36.2, Berlin and New York: Walter de Gruyter, pp. 862-909.

Drabkin, Israel E. (1935), 'Aristotle, *De Caelo* 2.6, 288a, 22', *The Classical Weekly* 29, 93-6.

Drossart Lulofs, H.J. (1965), *Nicolaus Damascenus. On the Philosophy of Aristotle* (Philosophy Antiqua vol. 13), Leiden: E.J. Brill.

Failler, Albert (ed. and trans.) (1999), *Georges Pachymérès, Relations Historiques III, Livres VII-IX*, Paris: Institut Français d'Etudes Byzantines.

Federspiel, Michel (1992), 'Sur le mouvement des projectiles (Aristote, *Du ciel*, 288a22)', *Revue des Études Anciennes* 94, 337-45.

Federspiel, Michel (1995), 'Notes sur le traité aristotélicien *Du Ciel*', *Revue des Études Anciennes* 97, 505-16.

Festugière, A.J. (trans.) (1966-1968), *Proclus, Commentaire sur le Timée*, 3 vols, Paris: Libraire Philosophique J. Vrin.

Frede, Michael (1974), *Die Stoische Logik* (Abhandlungen der Akademie der Wissenschaften in Göttingen: Philologisch-Historische Klasse, Dritte Folge 88), Göttingen: Vandenhoeck & Ruprecht.

Gantz, Timothy (1993), *Early Greek Myth*, Baltimore and London: Johns Hopkins University Press.

Genequand, Charles (ed. and trans.) (2001), *Alexander of Aphrodisias on the Cosmos* (Islamic Philosophy, Theology, and Science, Texts and Studies 44), Leiden: E.J. Brill.

Gershenson, Daniel D. and Greenberg, Daniel A. (1964), *Anaxagoras and the Birth of Physics*, New York, London and Toronto: Blaisdell Publishing Company.

Goldstein, Bernard R. (1967), *The Arabic Version of Ptolemy's Planetary Hypotheses* (Transactions of the American Philosophical Society n.s. 57.4), Philadelphia: The American Philosophical Society.

Gottschalk, H.B. (1980), *Heraclides of Pontus*, Oxford: Clarendon Press.

Goulet, Richard (ed.) (1989-), *Dictionnaire des Philosophes Antiques*, Paris: Éditions du Centre National de la Recherche Scientifique.

Greene, William Chase (1948), *Scholia Platonica* (Philological Monographs Published by the American Philological Association 7), Haverford: American Philological Association.

Guthrie, W.K.C. (1962-1981), *A History of Greek Philosophy*, 6 vols, Cambridge: Cambridge University Press.

Hadot, Ilsetraut (1978), *Le Problème du Néoplatonisme Alexandrin: Hiéroclès et Simplicius*, Paris: Études Augustiniennes.

Hadot, Ilsetraut (ed.) (1987), *Simplicius, sa Vie, son Oeuvre, sa Survie*, Berlin and New York: Walter de Gruyter.

Hadot, Ilsetraut (ed.) (1996), *Simplicius, Commentaire sur le Manuel d'Épictète* (Philosophia Antiqua 66), Leiden: E.J. Brill.

Heath, Thomas (1913), *Aristarchus of Samos*, Oxford: Clarendon Press.

Heiberg, J.L. (1892), 'Handschriftliches zum Commentar des Simplicius zu Aristoteles de caelo', *Sitzungsberichte der königlich preussischen Akademie der Wissenschaften zu Berlin*, Berlin: Georg Reimer, pp. 59-66.

Heiberg, J.L. (ed.) (1898-), *Claudii Ptolemaei Opera Quae Exstant Omnia*, 3 vols, Leipzig: Teubner.

Heiberg, J.L. and Menge, Hermann (ed. and trans.) (1883-1916), *Euclidis Opera Omnia*, Leipzig: Teubner.

Henry, Paul and Schwyzer, Hans-Rudolf (eds) (1951), *Plotini Opera*, vol. 1, Paris: Desclée de Brouwer, and Bruxelles: L'Édition Universelle.

Hoffmann, Philippe (1981), *Recherches sur la Tradition Manuscrite du Commentaire de Simplicius au De Caelo d'Aristote*, Thèse présentée à l'Université de PARIS IV pour le Doctorat de Troisième Cycle.

Hoffmann, Philippe (1987), 'Simplicius' Polemics', in Sorabji (1987), pp. 57-83.

Hultsch, Fridericus (ed.) (1876-1878), *Pappi Alexandri Collectionis Quae Supersunt*, 3 vols, Berlin: Weidmann.

Joachim, Harold H. (1922), *Aristotle on Coming-To-Be & Passing-Away*, Oxford: Clarendon Press.

Kahn, Charles H. (1979), *The Art and Thought of Heraclitus*, Cambridge: Cambridge University Press.

Kern, Otto (1922), *Orphicorum Fragmenta*, Berlin: Weidmann.

Kidd, Douglas (ed. and trans.) (1997), Aratus, *Phaenomena* (Cambridge Classical Texts and Commentaries 34), Cambridge: Cambridge University Press.

Kirk, G.S. (1954), *Heraclitus: The Cosmic Fragments*, Cambridge: Cambridge University Press.

Kissling, Robert Christian (1922), 'The OXHMA-PNEUMA of the Neo-Platonists and the *de Insomniis* of Synesius of Cyrene', *American Journal of Philology* 43, 318-30.

Lasserre, François (ed. and trans.) (1966), *Die Fragmente des Eudoxos von Knidos*, Berlin: Walter de Gruyter.

Long, A.A. (1974), 'Empedocles' Cosmic Cycle in the Sixties', in Mourelatos (1974), pp. 397-425.

Long, A.A. and Sedley, D.N. (1987), *The Hellenistic Philosophers*, 2 vols, Cambridge: Cambridge University Press.

Louis, Pierre (ed. and trans.) (1973), *Marche des Animaux; Mouvement des Animaux*, Paris: Belles Lettres.

Majercik, Ruth (1989), *The Chaldean Oracles*, Leiden: E.J. Brill.

Manitius, Carolus (ed. and trans.) (1909), *Proclis Diadochi Hypotyposis Astronomicarum Positionum*, Stuttgart: Teubner.

Marcovich, Miroslav (ed. and trans.) (1967), *Heraclitus*, Merida: Los Andes University Press.

Marcovich, Miroslav (ed.) (1999), *Diogenis Laertii Vitae Philosophorum*, Stuttgart and Leipzig: Teubner.

Merlan, Philip (1935), 'Ein Simplikios-Zitat bei Pseudo-Alexandros und ein Plotinos-Zitat bei Simplikios', *Rheinisches Museum* 84, 154-60. Reprinted in Merlan (1976), pp. 412-18.

Merlan, Philip (1943), 'Plotinus, *Enneads* 2.2', *Transactions of the American Philological Association* 74, 179-91. Reprinted in Merlan (1976), pp. 396-408.

Merlan, Philip (1968), 'Ammonius Hermiae, Zacharias Scholasticus and Boethius', *Greek, Roman, and Byzantine Studies* 9, 193-203. Reprinted in Merlan (1976), pp. 431-41.

Merlan, Philip (1976), *Kleine Philosophische Schriften*, Hildesheim and New York: Georg Olms.

Mioni, Elpidio (1985), *Bibliothecae Divi Marci Venetiarum, Codices Graeci Manuscripti*, vol. 2 'Thesaurus Antiquus Codices 300-625', Rome: Istituto Poligrafico e Zecca Dello Stato.

Moraux, Paul (1954), 'Notes sur la tradition indirecte du "De caelo" d'Aristote', *Hermes* 82, 145-82.

Moraux, Paul (1973), *Der Aristotelismus bei den Griechen*, vol. 1, Berlin and New York: Walter de Gruyter.

Moraux, Paul (1984), *Der Aristotelismus bei den Griechen*, vol. 2, Berlin and New York: Walter de Gruyter.

Moraux, Paul (2001), *Der Aristotelismus bei den Griechen*, vol. 3, Berlin and New York: Walter de Gruyter.

Mueller, Ian (2000), 'Plato's Geometrical Chemistry and its Exegesis in Antiquity', in Suppes, Moravcsik, and Mendell (2000), pp. 159-76.

Müller, Wilhelm (1953), 'Das isoperimetrische Problem im Altertum', *Sudhoffs Archiv für Geschichte der Medizin und der Naturwissenschaften* 37, 39-71.

Neugebauer, Otto (1975), *A History of Ancient Mathematical Astronomy*, 3 vols, Berlin: Springer-Verlag.

Obertello, Luca (ed. and trans.) (1969), *A. M. Severino Boezio De Hypotheticis Syllogismis*, Brescia: Paideia.

O'Brien, D. (1969), *Empedocles' Cosmic Cycle*, Cambridge: Cambridge University Press.

Pedersen, Olaf (1974), *A Survey of the Almagest* (Acta Historica Scientiarum Naturalium et Medicinalium 30), Odense: Odense Universitetsforlag.

Peyron, Amedeo (1810), *Empedoclis et Parmenidis Fragmenta*, Leipzig: I.A.G. Weigl.

Praechter, Karl (1924), 'Simpl. in Aristote de caelo p. 370,29 ff. H.', *Hermes* 59, 118-19.

Rome, A. (ed.) (1936), *Commentaires de Pappus et de Théon d'Alexandrie sur L'Almageste*, vol. 2 (*Studi e Testi* 72), Vatican City: Biblioteca Apostolica Vaticana.

Rose, Valentinus (1886), *Aristotelis Qui Ferebantur Librorum Fragmenta*, 3rd edition, Leipzig: Teubner.

Ross, W.D. (1953), *Aristotle's Metaphysics. A Revised Text with Introduction and Commentary*, 2 vols, Oxford: Clarendon Press.

Ruelle, Charles Emile (ed.) (1889), *Damascii Successoris, Dubitationes et Solutiones de Primis Principiis, in Platonis Parmenidem*, 2 parts, Paris: C. Klincksieck.

Sharples, R.W. (1983), 'The Unmoved Mover and the Motion of the Heavens in Alexander of Aphrodisias', *Apeiron* 17, 62-6.

Sharples, R.W. (1987), 'Alexander of Aphrodisias: Scholasticsm and Innovation', in Wolfgang Haase (ed.), *Aufstieg und Niedergang der Römischen Welt*, II 36.2, Berlin and New York: Walter de Gruyter, pp. 1176-243.

Sharples, R.W. (trans.) (1992), *Alexander of Aphrodisias, Quaestiones 1.1-2.15*, London: Duckworth.

Sorabji, Richard (ed.) (1987), *Philoponus and the Rejection of Aristotelian Science*, London: Duckworth.

Suppes, Patrick, Moravcsik, Julius M. and Mendell, Henry (eds) (2000), *Ancient and Medieval Traditions in the Exact Sciences*, Stanford: CSLI Publications.

Thiel, Helmut van (ed.) (1991), *Homeri Odyssea*, Hildesheim and New York: G. Olms.

Valk, Marchinus van der (ed.) (1987), *Eustathii Archiepiscopi Thessalonicensis Commentarii ad Homeri Iliadem Pertinentes*, vol. 4, Leiden: E.J. Brill.

Wallis, R.T. (1995), *Neoplatonism*, 2nd edition, Indianapolis: Hackett.

Wartelle, André (1963), *Inventaire des Manuscrits Grecs d'Aristote et de ses Commentateurs*, Paris: Belles Lettres.

West, M.L. (ed.) (1966), *Hesiod, Theogony*, Oxford: Clarendon Press.

West, M.L. (1983), *The Orphic Poems*, Oxford: Clarendon Press.

West, M.L. (ed.) (1998), *Homeri Ilias*, vol. 1, Stuttgart and Leipzig: Teubner.

Westerink, L.G. (ed.) (1956), *Olympiodorus, Commentary on the First Alcibiades of Plato*, Amsterdam: North Holland.

Westerink, L.G. (1977), *The Greek Commentaries on Plato's Phaedo*, vol. 2 (Verhandelingen der Koninklijke Nederlandse Akademie van Wetenschappen, Afd. Letterkunde n.s. 93), Amsterdam: North Holland.

Wright, M.R. (1981), *Empedocles: The Extant Fragments*, New Haven and London: Yale University Press.

Textual Questions

(a) Textual suggestions

Listed here are places where I have translated a text different from the one printed by Heiberg. For reasons which I hope I have made clear in the Introduction I prefer to call these deviations suggestions rather than emendations. In many cases notes on the lines in the translation provide more information.

370,7 For *athananton* read *athanaton* (misprint).
370,31 For *diablepomenoi* read *diabeblêmenoi* (a suggestion of Praechter (1924)).
371,4 For *genomenon* read *genomenên* with A.
371,24 For *tan* read *tas* (misprint).
372,11 For *aphôrismena* read *aphôrismenon* with F and Karsten.
375,19-20 Heiberg prints
hôs gar hupezôkôs tis, ou mên noeros, diakrinei kata to logion pur prôton kai pur heteron speudonta migênai;
This should probably be:
'hôs gar hupezôkôs tis humên noeros diakrinei' (*kata to logion*) *'pur prôton kai pur heteron speudonta migênai;'*
See the note ad loc.
377,11-12 For *surphetên logikôn te kai alogôn energeian* read *summikton phusin logikôn te kai alogôn energeiôn* with Karsten.
377,18 For *ampsukhai* read *anapsukhai* with F and Karsten.
378,19 For *sômatos* read *sômatôn.*
379,29 For *tois* read *autois* (Heiberg's suggestion).
380,35 For *tên psukhên* read *tês psukhês* with F and Karsten.
380,36 For *phuseôs* read *phusis* with Simplicius *in Phys.* (*CAG*, vol. 10) 1219,6.
382,19 For *dioti* read *dia ti* with F and Karsten.
385,34 For *triton* read *pempton.*
391,18 Bracket *monon.*
394,23 For *tên <tôn> planômenôn sphairan* read *tên planômenên sphairan* with F and Karsten.
394,35 For *opsei* read *opsetai* with Karsten.
395,26 For *diaphora* read *diaphoron* with F and Karsten.
400,7 For *akinêton* read *kinêton* or *aei kinêton.*
401,9 For *diakosmia tade* read *diakosmia ta tessara* with F, a reading which Heiberg thinks may be right.
404,27 For *tôn* read *hê tôn* with F and Karsten.
406,24 For *dêlon* read *dêlon oun*, a suggestion of Heiberg.
407,1 For *genêton* read *gennêsan* with Karsten and a correction of Bessarion.
408,23 For *ei* read *ei hê* with F and Karsten.
411,5 For *ta* read *tas* with Karsten (probably a misprint in Heiberg).
413,26 For *tên tou* read *to tên* with F and Karsten.

417,27 Bracket *ti*.
419,34 Omit *tôi* with K and Karsten.
420,8 For *ouanon* read *ouranon* (misprint).
420,19 Punctuate with a ';' rather than a '.' (with Karsten).
421,18 For *dêmiourgousi* read *dêmiourgei* with F and Karsten; for *tetagmenoi* read *tetagmena* with F and Karsten.
423,31 For *autôi* read *autois*.
426,16 Bracket *to kinoun*.
428,19 For *tên* read *tên autou* with F (Karsten has *tên hautou*).
428, 22 For the first *têi* read *gê* with F, b, and Karsten; for the second read *pur* with F, b, and Karsten.
428,31 For *diatithetai* read *kineitai*.
428,34 For *homalotêtos* read *kath' holotêta* with Karsten (F has *holotêta*).
433,28 For *autôi* read *auto* with A, the first hand of F, b, and Karsten.
436,17 For the second *to* read *kai* with a correction of Bessarion, b, and Karsten (F has *to*).
436,27 Put a period after *Alexandrou*. For *legein* read *legein hoti* with F and Karsten, and eliminate the semicolon in line 28.
440,10 For *megethos* read *megethos tou sômatos* with F and Karsten.
441,3 For *sômatikas* I suspect one should read *asômatikas*.
447,21 For *autôn* read *auton* with F and Karsten.
448,33 After *megethous* insert <*tôn kuklôn*> with Karsten.
449,26 For the two *ta* read *to* with F and Karsten.
449,27 Bracket *to*.
450,25 For *têi* read *têi autêi* with F and Karsten.
451,15 For *sunkinoumenoi* read *kinoumenoi* with Karsten (F has *hoi sunkinoumenoi*).
452,5 For *hoi analogon ekhontes* read *eisin hous legei ho Aristotelês analogon ekhontas* with F and Karsten.
453,9 For *parapempontes* read *parapleontes* with F and Karsten.
454,6 For *êkouen* read *êkousen* with F and Karsten.
458,27 For *kat' auton* read *kath' hauton* with A.
459,2 For *tôn isoperimetrôn khôriôn* read *tôn ison khôrion periekhousôn grammôn*.
463,6 For *homônuma* read *homônumôs* with D, E, F, and Karsten.
465,10 For *kompsôs* read *kompsôs auton* with D, E, F, and Karsten.
468,1 For *enüparkhontos* read *ekei huparkhontos* with D, F, a correction in E, and Karsten.
468,15 For *hupo* read *hupo tou Aristotelous kai* with E, F, and Karsten (D has *hupo tou Aristotelous kai ta*).
468,16 For *metabatika* read *metabatikôs* with D, E, F, and Karsten.
469,7 For *to autoeides autou kai ouranion okhêma* read *tou augoeidous autou kai ouraniou okhêmatos* (A, D, E, F, and Karsten all have *augoeides*).
469,23 For *rhêteon* read *rhêteon isôs* with D, E, F, and Karsten.

(b) Simplicius' citations of *On the Heavens* 2.1-9

Here I bring together places where the text of a citation by Simplicius of a passage from *De Caelo* 2.1-9 as printed by Heiberg differs from Moraux's text of Aristotle. I also indicate what is printed by Karsten and (where Heiberg provides the information) what appears in F. In general Heiberg's text reproduces A. I have paid no attention to the numerous differences regarding elision (e.g. *de* vs. *d'*) or minor variations in spelling (e.g. *hauton* vs. *heauton* or *teleiotaton* vs. *teleôtaton*).

Moraux	Heiberg	Karsten
284b16 *panta*	386,25 omit	*panta* (also F)
284b5 *to nun*	382,20 *to nun*	*ta nun* (also F)
284b23 *gar*	386,28 *de*	*gar* (also F)
284b24 *pasas*	386,29 omit	*pasas* (also F)
285a33 *tôn moriôn hapantôn*	389,8 *pantôn tôn moriôn*	*tôn moriôn hapantôn*
286a22 *mên*	399,29 omit	*mên*
286a35 *einai kata phusin tên kinêsin*	402,11 *kata phusin tên kinêsin einai*	286a35 *einai kata phusin tên kinêsin*
287a23 *phoras*	409,31 *periphoras*	*phoras* (and at 410,11)
287a28 *aph' hautou eph' hauto*	412,11 *apo tou autou epi to auto*	*apo tou autou epi to auto*
	413,10 *apo tou autou epi to auto*	*apo tautou epi to auto*
	413,14 *apo tou autou epi to auto*	*apo tautou epi to auto*
	413,16 *apo tou autou epi to auto*	*apo tou autou epi to auto*
	414,5 *aph' heautou eph' heauto*	*apo tautou epi tauton*
289a3 *tôi autôi te*	34,14 *tôi autôi*	*tautôi t'*
289a25 *auta*	439,21 *hautai*	*hautai*
289b15 *gar*	447,26 omit	omit
289b15 *all'*	447,26 *alla kai*	*alla kai*
289b25 *kai ouk*	447,4 *ouk*	*kouk* (F has *kai ouk*)
289b27 *to apo tukhês*	447,5 *apo tukhês*	*to apo tukhês; apo tukhês* (also F)
290a6 *sunekhes on*	451,8 *sunekhes*	*sunekhes on* (also F)
290a8 *eipein*	477,10 *eiper*	*eiper*
290a12 *men gar*	455,14 *gar*	*men gar*
	456,16 *gar*	*gar*
	456,31 *gar*	
290a13 *ton topon*	455,15 *topon*	*ton topon*
290b3 *katekhoi*	455,19 *katekhei*	*katekhoi*
	459,1 *katekhoi*	*katekhoi* (F has *katekhei*)
	478,24 *katekhoi*	*katekhoi*
290b24 *dokei*	464,27 *edokei*	*edokei*
290b32 *legein*	465,18 *luein*	*luein*

(c) Simplicius' citations of other texts

Here I bring together places where the text of a citation by Simplicius of a passage from a work other than *De Caelo* as printed by Heiberg differs from the text of a standard edition of the work. I also indicate what is printed by Karsten and (where Heiberg provides the information) what appears in F. In general Heiberg's text reproduces A. I have paid no attention to the numerous differences regarding elision (e.g. *de* vs. *d')* or minor variations in spelling (e.g. *hauton* vs. *heauton* or *teleiotaton* vs. *teleôtaton*).

Aristotle, *De Incessu Animalium*

Louis (1973)	Heiberg	Karsten
705b10 *to opisthen*	384,14 *opisthen*	*to opisthen* (also F)
705b17 *homoiôs*	384,18 *kai homoiôs*	*homoiôs*

Homer, *Iliad*

West (1998)	Heiberg	Karsten
12,239 *dekhi' iôsi*	392,7 *dekhia esti*	*dekhi' iôsi* (also F)

Plato, *Laws*

Oeuv. Comp. vol. 12.1	Heiberg	Karsten
895A6 *pôs*	389,31 omit	*pôs*

Plato, *Politicus*

Oeuv. Comp. vol. 9.1	Heiberg	Karsten
270A8 *epi mikrotatou bainon podos*	459,6 *epi smikrotatou podos aei bainon*	*epi smikrotatou podos aei bainon*

Plato, *Timaeus*

Oeuv. Comp. vol. 10	Heiberg	Karsten
33B4 *hoposa*	408,34 *hosa*	*hosa*
33B5 *tas*	408,34 omit	*tas* (also F)
	418,9 *tas*	*tas*
33B6 *etorneusato*	408,35 *etornôsato*	*etorneusato* (F has *etornôsato*)
	418,10 *etornôsato*	*etornôsato*
40A3 *ek puros*	436,2 *ek tou puros*	*ek puros* (also F)
40A3 *apêrgazeto*	454,30 *apeirgasato*	*apêrgazeto* (F has *apeirgasato*)
40A7 *einai*	455,1 *einai kai*	*einai*
40A8 *kata tauta peri tôn autôn aei ta auta heautôi dianooumenôi*	455,2 *kata tauto dinoumenôi*	*kata tauta peri tôn autôn aei tauto heautôi dianooumenôi* (F has *kata tauta peri tôn autôn aei tauth' heautôi dinoumenon*)
40B3 *hestos*	455,4 *hestôs*	*hestôs*
40B3 *autôn*	455,4 *kai autôn*	*autôn*
40B5 *zôia theia*	455,6 *theia*	*zôia thei'* (F has *theia*)
40B6 *aidia kai kata tauta*	455,6 *dia tauta*	*aidia kai kata taut'*

(d) Lemmas

Here I bring together places where the text in a lemma printed by Heiberg differs from Moraux's text of Aristotle. I also indicate what is printed by Karsten and (where Heiberg provides the information) what appears in F. In general Heiberg's text reproduces A. I have paid no attention to the numerous differences regarding elision (e.g. *de* vs. *d'*) or minor variations in spelling (e.g. *hauton* vs. *heauton* or *teleiotaton* vs. *teleôtaton*). I should perhaps note that the lemmas in Heiberg and Karsten generally give only the first and last few words of a passage, and so represent less than 10 per cent of the text of *De Caelo*.

Moraux	Heiberg	Karsten
284b5 *to nun*	374,2 *ta nun*	*ta nun*
286a8 *estin*	395,27 *de*	*estin*
287b14 *BEG*	415,33 *hê BEG*	*BEG* (also F)
287b21 *sustoikhôn*	417,16 *stoikheiôn*	*stoikheiôn*
288a27 *d' epei*	425,27 *epei*	*epei* (also F)
288a27 *to*	425,27 omit	*to* (also F)
289a10 *hêmin*	434,31 omit	*hêmin* (also F)

289a11 *astrôn*	435,10 *asterôn*	*astrôn*
289b27 *to apo tukhês*	444,17 *tês tukhês*	*to apo tukhês* (F has *apo tukhês*)
289b27 *kata ta megethê tôn kuklôn*	448,10 omit	*kata ta megethê tôn kuklôn*
289b30 *Epei*	449,3 *Epeidê*	*Epei* (also F)
290a29 *kinoito*	456,29 *kinointo*	*kinoito*
	458,7 *kinoito*	*kinointo* (also F)
290b6 *kinêtikois*	478,25 *kinoumenois*	*kinêtikois*
290b9 *ta d' alla* [*astra*]	455,22 *ta de alla astra*	*talla de astra*
	460,5 *ta de alla ourania sômata, toutesti ta astra*	*ta de alla ourania sômata, toutesti ta astra*

English-Greek Glossary

This glossary gives standard Greek equivalents for many nouns, verbs, adjectives, adverbs, and a few prepositions in the translation. Most Greek words which occur only once are omitted, as are many words which have been given a variety of translations (e.g. *logos* and *apodidonai*) and certain very common words (e.g. *einai*, *ekhein*, and *legein*). The reader will get a better sense of the range of a Greek word by looking at the Greek-English Index for the word and ones closely related to it.

absurd: *atopos*
accelerate: *epiteinai*; acceleration: *epitasis*
accept: *apodekhesthai*, *dekhesthai*
accessible: *prokheiros*
account (v.): *apoligizesthai*
accuse: *enkalein*
act: *energein*
action: *praxis*
active: *energêtikos*, *poiêtikos*
activity: *energeia*
actualisation: *entelekheia*
actualise: *energein*
add: *epagein*, *prostithenai*
addition: *prosthêkê*, *prosthesis*
affected, be: *paskhein*
affection: *pathos*
affirmation: *kataphasis*
aggregate (v.): *sunkrinein*; aggregation: *sunkrisis*
agree: *homologein*, *sugkhôrein*
aim (v.): *teinein*
air: *aêr*
aitherial: *aitherios*, *aitherôdês*
alter: *metagraphthai*
alteration: *alloiôsis*
alternating: *enallax*
analogous: *analogon*
analogy: *analogia*
ancient: *arkhaios*
angle: *gônia*
animal: *zôion*
announce: *epangellesthai*
anomalous: *atopos*
answer (n.): *lusis*; answer (v.): *apokrinesthai*, *luein*
antecedent: *hêgoumenon*
antithesis: *antithesis*
apart: *khôris*
appearance: *phantasia*
apprehend: *antilambanesthai*, *katalambanein*
appropriate: *oikeios*
arc: *periphereia*
area: *embadon*, *khôrion*
argue: *sullogizesthai*; argue (for): *kataskeuazein*
argument: *epikheirêma*
arrangement: *diathesis*
art: *tekhnê*
artificial: *tekhnikos*
artist: *tekhnitês*
ask: *erôtan*
assign: *tattein*
assume: *lambanein*
astronomer: *astronomos*
attack (v.): *diaballein*
attention: *epistasis*
authoritative: *kurios*
axiom: *axiôma*
axis: *axôn*

back (adv., adj.): *opisthen*, *opisthios*
backward: *opisô*
base: *basis*
beauty: *kallos*

begin: *arkhein*
belief: *pistis*
believability: *pistis*
belong: *huparkhein*
below: *katô*
besouled: *empsukhos*
best: *aristos*
better: *kreittôn*
bind together: *sunekhein*
black (adj.): *melas*
blacksmith: *khalkotupos*
blessedness: *makariotês*
boat: *naus*
body: *sôma*
book: *biblion*
breadth: *platos*
brief: *brakhus, suntomos*
bright: *lampros, stilpnos*
bring together: *sunairein*
bulk: *onkos*

call (v.): *kalein*
capacity: *dunamis*
carry around: *sumperiagein, semperipherein*
carry out: *proagein*
cause: *aitia, aition*
cautious: *asphalês*
cease: *pauesthai*
censure: *aitiasthai, enkalein, memphesthai*
centre: *mesos*
central point: *kentros*
cessation: *paula*
chafed, be: *paratribesthai, parektribesthai*
chance: *tukhê, tukhôn*
change (n.): *metabolê*; change (v.): *metabainein, metaballein, parallattein, ameibein*; change of place: *metabasis*; change position: *methistanai*; changing place: *metabatikos*
characterise: *eidopoein, kharaktêrizein*
circle: *kuklos*
circular: *enkuklios, kuklikos*; circular (of an argument): *diallêlos*
circular motion: *kuklophoria, enkuklios kinêsis* (or *phora*), *kuklikê kinêsis* (or *phora*), *kuklôi kinêsis* (or *phora*), and *en kuklôi kinêsis* (or *phora*)
circumference: *periphereia*
clear: *dêlos, enargês, saphês*
clever: *kompsos, perittos*
clod of earth: *bôlos*
close: *engus*
cloud: *nephelê, nephos*
cold (adj.): *psukhros*; coldness: *psukhrotês, psuxis*
collection: *sunagôgê*
colour: *khroa, khrôma*
column: *sustoikhia*
combination, astrological: *sunkrasis*
come to be: *gignesthai*; coming to be (adj.): *genêtos*; coming to be (n.): *genesis*
common: *koinos, sunêthês*
compare: *paraballein, sumballein*
comparison: *sunkrisis*
complete (adj.): *pantelês*; *teleios*; complete (v.): *teleioun*; completely: *teleôs*; completeness: *teleiotês*
completing: *teleiôtikos*
composite (adj.): *miktos, sunthetos*
compounded, be: *sunkeisthai*
compress: *pilein, sumpilein, sunthlibein, puknoun*
compression: *pilêsis*
concave: *koilos*
conception: *ennoia, epinoia, hupolêpsis*
conclude: *sumperainein*
conclusion: *sumperasma*
condition: *katastasis, moira*
cone: *kônos*
configuration: *skhêmatismos*
confirm: *pistousthai*
conjunction, be in: *sunodeuein*
connect: *sunaptein*
consequence: *akolouthia*
consequent: *hepomenos*
consider: *episkeptesthai*
constrain: *anankazein*
construction: *kataskeuê, sustasis*; construction (of an argument): *sunagôgê*
contact (n.): *haphê*
contain: *periekhein*
container: *angeion*
contest (v.): *hupantan*
continue: *diatelein*
continuity: *sunekheia*
continuous: *sunekhês*
contrary motion: *antikinêsis*

contrast (n.): *antidiastolê*; contrast (v.): *antidiastellein*
coordinate (adj.): *sustoikhos*; coordinate (v.): *suntattein*
corporeal: *sômatikos, sômatoeidês*
correctly: *kalôs, orthôs*
course: *hodos*
curvilinear: *peripherogrammos*
cut: *temnein*
cylinder: *kulindros*

day: *hêmera*
decelerate: *anienai*
deceleration: *anesis*
degree (geometric): *moira*
delimit: *hôrizein*
demiurge: *dêmiourgos*
demonstrate: *apodeiknunai*; demonstrate already: *proapodeiknunai*; demonstrate in addition: *prosapodeiknunai*
demonstration: *apodeixis*
demonstrative: *apodeiktikos*
dense: *pakhumerês, pakhus, puknos*; denseness: *puknotês*
density: *puknôsis*
deny: *apophanai*
depth: *bathos*
describe: *paristanei*
desire (n.): *horexis, hormê*
destroy: *diaphtheirein, phtheirein, sunanairein*; destroyed: *phthartos*
destructible: *phthartos*
detached: *apêrtêmenos*
determine: *aphorizein, diorizein, hôrizein*
diameter: *diametros*
differ: *diapherein*
difference: *diaphora*
different: *diaphoros*
differentia: *diaphora*
difficult: *aporos*; difficulty: *aporia, duskhereia*
dimension: *diastasis*
diminish: *meioun*
direct (v.): *teinein*; direct (adj.): *prosêkhês*
direction: *skopos*
discover: *heuriskein*
discuss: *dialegesthai*
disposition: *diathesis*
dissimilar: *anomoios*
dissolve: *dialuein, epiluein, luein*
distance: *apostasis, apostêma, diastasis, diastêma*
distinction: *diakrisis*
distinguish: *diakrinai, diorizein*
divide: *diairein, diakrinai, dialambanein*
divine: *theios*
division: *diairesis, diakrisis*
doctrine: *dogma*
dominate: *epikratein*
domination: *kratêsis*
down, downward: *katô*
draw: *agein, graphein*
duration: *aiôn*

earlier: *palai*
earth: *gê*
earthy: *geêros, gêinos*
east (adj.): *anatolikos*; east (n.): *anatolê*
easy: *prokheiros*
eclipse: *ekleipsis*
effortless: *aponos*
element: *stoikheion*
eliminate: *anairein, aphairein*
emend: *metagraphthai*
emit: *ekdidonai*
end (n.): *telos*
endure: *menein*
entire: *holos*; entirety: *holotês*
equal: *isos, homoios*
equator: *isêmerinos* (*kuklos*)
equinox: *isêmeria*
establish: *enidruein*
eternal: *aidios*
eternity: *aidiotês*
everywhere: *pantakhou, pantakhothen*
evidence: *tekmêrion*
evident: *phaneros*
example: *paradeigma*
excessive: *perittos*
exist: *huparkhein*; exist after: *methuparkhein*; exist before or in a prior way: *prouparkhein*; exist together: *sunuparkhein*
existence: *hupostasis*
explain: *didaskein*
explanation: *aitia, aition*
extend: *ekballein, ekteinein, prosekballein*
extension: *diastêma*; (temporal) extension: *paratasis*

extrusion: *exokhê*
eye: *ophthalmos*

face: *prosôpon*
fall behind: *hupoleipein*
fall out: *sumpiptein*
familiar: *gnôrimos*
familiarity: *suntrophia*
fast: *takhus*; faster: *thattôn*
fasten: *endein*
fate: *moira*
father: *patêr*
fiction: *plasma*
fiery: *purinos*
figure: *skhêma*
fill: *plêroun*, *sumplêroun*
final: *loipos*
find: *heuriskein*
fine: *leptomerês*, *leptos*
fineness: *leptotêtos*
finger: *daktulos*
finish (up): *sumperainein*
finishing point: *telos*
finite: *peperasmenos*
fire: *pur*
firm: *bebaios*
fit (v.): *harmozein*
fitting: *prepôn*
fixed: *aplanês*
flow: *surrein*
follow: *akolouthein*, *hepesthai*
foot: *pous*
force: *bia*; forced: *biaios*
forearm: *pêkhus*
foresee: *pronoein*
form (n.): *eidos*
forward: *prosô*
freedom: *rhastônê*
front: *emprosthen*, *prosthen*
function: *ergon*

general: *koinos*
generate: *gennan*
genus: *genos*
give: *didonai*; give shape: *morphoun*
go together: *sunodeuein*
god: *theos*
good: *agathos*
governance: *dioikêsis*
great: *megas*
grow: *auxanein*
habitual: *sunêthês*
habituation: *sunêtheia*

hand: *kheir*
harmonious: *harmonikos*, *sumphônos*
harmonise: *sunaidein*, *sunarmozein*; harmonised: *enarmonios*
harmony: *harmonia*, *sumphônia*
head: *kephalê*
hear: *akouein*
heat (n.): *thermotês*; heat (v.): *thermainein*, *ekthermainein*
heaven: *ouranos*; heavenly; *ouranios*
heavy: *barus*; heaviness: *barutês*
hemisphere: *hêmisphairion*
high: *meteôros*; highest form: *akrotês*
high pitch: *oxutês*
hold: *huparkhein*
hollow (adj.): *koilos*, *pharangôdês*; hollow (n.): *koilotês*
homoiomerous: *homoiomerês*
honourable: *timios*
horizon: (*hô*) *horizôn*
horse: *hippos*
hot: *thermos*
hotness: *thermotês*
human (being): *anthrôpos*
hypothesis: *hupothesis*
hypothesise: *hupotithenai*
hypothetical: *hupothetikos*

idea: *epibolê*
illuminate: *phôtizein*; illumination: *phôtismos*
image: *eidôlon*, *eikôn*, *emphasis*
imagination: *phantasia*
imagine: *phantazein*
immaterial: *aülos*
immortal: *athanatos*
immortality: *athanasia*
impassive: *apathês*
implausible: *apithanos*
important: *kurios*
impossible: *adunatos*
impression: *hupopsia*
in proportion: *analogon*
inclination: *rhopê*
incline: *rhepein*
incomplete: *atelês*
incorporeal: *asômatos*
incredible: *amêkhanos*
indestructible: *aphthartos*
indicate: *endeiknunai*, *sêmainein*
indirect(ly): *kata sumbebêkos*
induction: *epagôgê*
infer: *sullogizesthai*

inference: *sunagôgê*
infinite: *apeiros*
inhere: *enuparkhein*
inherent: *sumphuês, sumphutos*
inquire: *epizêtein*
inquiry: *skemma, zêtêma, zêtêsis*
inseparable: *akhôristos*
inside: *endon*
instrument: *organon*
intellect: *nous*
intellectual: *noeros*
intelligible: *noêtos*
intention: *pronoia*
intermediate: *mesos*
interpret: *exêgêsthai*
interpretation: *exêgêsis*
intersection: *sumbolê*
interval: *diastêma*
introduce: *prostithenai*
investigate: *skopein*
invisible: *aphanês*
irrational: *alogos*
isoperimetric: *isoperimetros*
issue: *problêma*

join: *epizeugnunai*

kind (n.): *genos*
kinetic: *kinêtikos*
know: *eidenai, epistasthai*
knowledge: *epistêmê, gnôsis*

labour (v.): *kamnein*
laborious: *epiponos*
large: *megas*
last (adj.): *eskhatos*
later: *husteros*
lead ball: *molibdis*
learn: *manthanein*
leave (behind): *apoleipein*; leave out: *paraleipein*
left (adj.): *aristeros, laios*
length: *mêkos*
lentil-shaped: *phakoeidês*
less: *elattôn, hêtton*
lie: *keisthai*; lie alongside: *parakeisthai*
life: *bios, zôê*
light (n.): *phôs*
limit (n.): *peras*; limit (v.): *aphorizein*
line: *grammê*
liquid (adj.): *hugros*
little: *brakhus, oligos*
live (v.): *oikein*
living thing: *zôion*
local: *topikos*
location: *oikêsis*
lower: *katô*
luminescence: *eklampsis*
lunar: *selêniakos*

magnitude: *megethos*
man: *anthrôpos*
manifest: *prodêlos*
manner: *tropos*
mass: *plêthos*
mast: *istos*
material: *enulos, hulikos*
matter: *hulê*
maximum (speed): *akmê*
mean (adj.): *mesos*; mean (v.): *sêmainein*
measure (n.): *metron*; measure (v.): *metrein*
melt: *têkesthai*
mental: *emphrôn*
meridian: *mesêmbria, mesêmbrinos kuklos*
method: *tropos*
middle: *mesos*
minimum (speed): *parakmê*
minor (adj.): *elattôn*
mirror: *katoptron*
missile: *belos*
mode: *tropos*
moist: *hugros*
monad: *monas*
moon: *selênê*
mortal: *epikêros*
motion: *kinêsis, phora*
motionless: *akinêtos*
mountain: *oros*
move (trans.): *kinein, pherein*; move (intrans.): *kineisthai, pheresthai*; move along with (intransitive): *sunkineisthai*; move in a circle: *kuklophereisthai*; moving in a circle: *kuklophorêtikos*
movable: *kinêtos*
multiple (adj.): *polueidês*
multiplicity, multitude: *plêthos*
myth: *muthos*; mythical: *muthikos*

natural: *autophuês, phusei, kata phusin, phusikos*
nature: *phusis*

near: *plêsios*; be or get near: *plêsiazein*
necessary: *anankaios*
necessity: *anankê*
net: *diktuon*
next: *ephexês, loipos*
node: *sundesmos*
non-uniform: *anômalos*; non-uniformity: *anômalia, anomoiotêta*
north (adj.): *boreios*
note (v.): *sêmeioun, ephistanai*
numbed, be: *narkan*
number: *arithmos*
nutritive: *threptikos*

object (v.): *enkalein, ephistanai, enistasthai*
objection: *enstasis*
oblique: *loxos*
observation: *têrêsis*
observe: *horan*
obvious(ly): *dêladê, dêlonoti*
occupy: *epekhein*
one's own: *oikeios*
opinion: *gnômê*
oppose: *antikeisthai*
opposite: *enantios*
opposition: *enantiôsis*
oracle: *manteia*
order (n.): *taxis*; order (v.): *tattein*
organ: *organon*
ought: *opheilein*
overflow: *ekkheisthai*

paradigm: *paradeigma*
parallel: *parallêlos*
part: *meros, moira, morion, tmêma*
participation: *methexis*
particular (adj.): *merikos, idios*
pass over: *parienai*
passage (in a text): *topos, khôrion*
passive: *pathêtikos*
path: *hodos*
patrimonial: *patrios*
pay attention: *ephistanai, ephistanein*
peculiar: *idios*
perceive: *aisthanesthai*
perceived, perceptible: *aisthêtos*
perception: *aisthêsis*
perfect (v.): *teleioun*
perimeter: *perimetros*
Peripatetic: *Peripatêtikos*
person (people): *anthrôpos*
philosopher: *philosophos*
physical: *phusikos*
pillar: *kiôn*
place (n.): *topos*
place (around): *peritithenai*
plane (figure): *epipedos*
planet: *planetês*
plant: *dendron*
Platonic: *Platonikos*
plausible: *pithanos*; make plausible: *pistousthai*
point: *sêmeion*
pointless(ly): *matên*
pole: *polos*
position: *thesis, topos*
positive condition: *hexis*
possible: *dunatos*
posterior: *husteros*
power: *dunamis*; loss of power: *adunamia*
practically: *skhedon*
precede: *proêgeisthai*
precise: *akribês*
precision: *akribeia*
predicate (v.): *katêgorein*
premiss: *protasis*
preparation: *paraskeuê*
present, be: *pareinai, prokeisthai*
preserve: *diasôzein, sôzein*
prevent: *kôluein*
primary, be: *proêgeisthai*
principle: *arkhê*
privation: *sterêsis*
problem: *problêma*
proceed: *proerkhesthai*
produce (v.): *parekhein, apotelein*; produce in: *endidonai*
product: *apotelesma*
productive: *oistikos, poiêtikos*
progress (v.): *proienai*; progress in a straight line: *euthuporesthai*
progression: *proödos*
project (v.): *rhiptein*; project upward: *anarriptein*
projecting: *proekhôn*
proof: *deixis*
proper: *oikeios*
proportion: *analogia*; proportional: *analogon*; proportionality: *analogia*
propose: *proballesthai, protithesthai*
prove: *deiknunai*

providence: *pronoia*
proximate: *prosêkhês*
prudence: *eulabeia*
purity: *katharotês*
purpose: *skopos*
purposively: *epitêdes*
push: *ôthein*
pushing (n.): *ôthismos*

quadrature: *tetragônismos*
quality: *poiotês*
quick: *takhus*

radius: *diastêma*
rare: *manos, spanios*
rational: *logikos*
ray: *aktis*
reach: *aphikneisthai*
reason: *aitia*; reasonable: *eikos, eulogos, kata logon*
recall: *hupomimnêskein*
receive: *apolambanein, dekhesthai*
receptacle: *angeion, hupodokhê*
recognise: *ephistanai, ephistanein*
revolution: *periphora*
recount: *historein*
rectilinear: *euthugrammos, euthugrammikos*
reduce: *anagein*
refer: *anapherein, mnêmoneuein*
reflect: *anaklan*
refute: *elegkhein*
region: *topos*
reject: *memphesthai*
relation: *skhesis*
relative: *pros ti*
remain: *diamenein, leipesthai, menein, perileipein*; remaining: *loipos*
remark: *ephistanai*
remind: *hupomimnêskein*
resemble: *eoikenai*
resistance: *antereisis, antitupia*
resolution: *lusis*
resolve: *luein*
resonance: *êkhos*
responsible: *aitios*
rest (n.): *êremia, monê*; rest (v.) *êremein*; rest period: *stasis*
restored, be: *apokathistasthai*; be restored simultaneously: *sunapokathistasthai*
revert: *epistrephesthai*
revolve: *peripheresthai*
right (adj.): *dexios*
rise (up): *anabainein, anatellein, aneinai, anekhein, anistanai*
rise together: *sunanatellein*
rising: *anatolê*
roll: *kuliesthai*
rolling (n.): *kulisis*
root: *rhiza*
rotate: *dineisthai*
rotation: *dinê, dinêsis, peristrophê*
round: *kukloterês, peripherês*
rubbing (n.): *tripsis*
run: *trekhein*
rushing (n.): *rhoizos*

same: *homoios*
science: *epistêmê*
scientist: *epistêmôn*
sea: *pelagos, thalassa*
see: *horan, idein*
seed: *sperma*
seem: *eoikenai*
self-moving: *autokinêtos*
sense, sense organ: *aisthêsis*
separable: *khôristos*
serviceable: *khrêsimos*
set on fire: *ekpuroun, exaptein*
setting (n.): *dusis*
shaking (n.): *tromos*
shape: *skhêma*
ship: *ploion*
short: *oligos*
shoulder: *ômos*
sickness: *nosos*
side: *plagios*
significance: *dunamis*
similar: *homoios*; similarity: *homoiotês*
simple: *haplos*
simultaneous restoration: *sunapokatastasis*
size: *megethos*
slander (v.): *diaballein*
slight: *brakhus*
slow: *bradus*; slowness: *bradutês*
small: *brakhus*; smaller: *elattôn*
smooth: *leios*
snow: *khiôn*
soft: *leios*
solar: *hêliakos*
solid (adj.): *stereos, steremnios*
solution: *lusis*

solve: *luein*
soul: *psukhê*; involving soul: *psukhikos*; soulless: *apsukhos*
sound (n.): *psophos*, *phônê*; to make sound: *psophein*
source: *pêgê*
southern: *mesêmbrinos*, *notios*
space: *khôra*
spacious: *polukhôros*
spark: *spinthêr*
spatial: *topikos*
spear: *doru*
specific: *idios*
specify: *diorizein*
speed: *takhos*, *takhutês*
sphere: *sphaira*
spherical: *sphairikos*, *sphairoeidês*; be made or become spherical: *sphairousthai*
spontaneous: *automatos*
spring (season): *ear*
square: *tetragônos*
stable: *monimos*
stade: *stadion*
stand: *istanai*
standing still (n.): *stasis*
star: *astêr*, *astron*
start (n.): *arkhê*; start (v.): *arkhein*
starting point: *arkhê*
sting ray: *narkê*
straight: *euthus*
strange: *atopos*
strive: *speudein*, *spoudazein*, *philoneikein*
strong: *enkratês*, *iskhuros*, *karteros*; stronger: *kreittôn*
structure: *sustasis*
study (n.): *theôria*
sublunar: *hupo selênên*
substance: *ousia*
substantial: *ousiôdês*
subtract: *aphairein*
subtraction: *huphairesis*
suffice: *arkein*
sufficient: *autarkês*, *hikanos*
suitability: *epitêdeiotêta*
suitable: *epitêdeios*; be suitable: *prosêkein*
summer: *theros*
superiority: *huperokhê*
suppose: *areskein*
supposition: *huponoia*
surface: *epiphaneia*
surprised, be: *thaumazein*
surprising: *thaumastos*
surrounding: *perix*
syllogism: *sullogismos*

take: *lambanein*, *paralambanein*
tall: *makros*
tangible: *haptos*
tear apart: *diaspan*
temporal: *khronikos*
termination: *teleutê*
text: *lexis*
theorem: *theôrêma*
theoretical: *theôrêtikos*
thesis: *thesis*
think: *boulesthai*, *noein*
thought: *epinoia*
thunder: *brontê*
time: *khronos*
today: *sêmeron*
topic: *kephalaion*
touch (v.): *ephaptesthai*, *haptesthai*
transcend: *exaireisthai*, *huperairein*; transcending: *epekeina*
transmit: *diabibazein*, *diapempein*, *diaporthmeuein*
transparent: *diaphanês*
traverse: *diexienai*, *diienai*
treatise: *pragmateia*
tree: *dendron*
tremble: *kradainesthai*
trouble (v.): *enokhlein*
true: *alêthês*
truth: *alêtheia*
try: *epikheirein*, *peiran*
turn (intransitive): *strephesthai*
turned out: *entornos*
twinkle: *stilbein*; twinkling (n.): *stilbêdôn*

unaltering: *analloiôtos*
unbelievable: *amêkhanos*
unbroken: *aklastos*
uncaused: *anaitios*
unceasing: *apaustos*
unchanging: *ametablêtos*, *ametapeistos*, *ametathetos*
unclear: *adêlos*
understand: *akouein*, *parakolouthein*
uniform: *homalês*; uniformity: *antanisôsis*, *homalotês*, *homoiotês*
unify: *henoun*
uninterrupted: *anekleiptos*

unique: *idios*
universal: *holikos, katholikos, katholou, koinos*
unmoving: *akinêtos*
unnatural: *para phusin*
unreasonable: *alogos, apeikotôs*
unserviceable: *akhrêstos*
use (n.): *khreia*; use (v.): *khrêsthai, paralambanein, proskhrêsthai, sugkhrêsthai*

variegated: *poikilos*
verify: *epideiknunai*
vertex: *koruphê*
view (v.): *blepein, horan*
visible: *horatos*
vision: *opsis*
void (adj.): *kenos*
vortex: *dinêsis*

wander: *planasthai*
water: *hudôr*
way: *tropos*

weaker: *kheirôn*
weakness: *astheneia*
weight: *baros, rhopê*
west (n.): *dusis, dusmê, dutikon*
white (adj.): *leukos*
whole: *holos*; wholeness: *holotês*
will: *boulêsis*
winter (adj.): *kheimerinos*; winter (n.): *kheimôn*
wisdom: *phronêsis*
wish (v.): *boulesthai*
witness (n.): *marturia, marturion*; bear witness: *marturein*
wonder (v.): *thaumazein*
wonderful: *thaumastos*
words: *lexis, rhêmata*
worse: *kheirôn*
worth (adj.): *axios*
write: *graphein*

year: *etos*

zenith: *koruphê*
zodiac: *ho zôidiakos kuklos, to zôidiakon*; sign of the zodiac: *zôidion*
zoophyte: *zôiophuton*

Greek-English Index

This index, which is based on Heiberg's text with my emendations, gives the English translations of many nouns, verbs, adjectives, adverbs, and a few prepositions used by Simplicius; certain very common words (e.g. *einai*, *ekhein*, and *legein*) and number words are omitted, as are words which only occur in quotations (or apparent quotations) of other authors (the exceptions to this are Alexander, Ptolemy, and Sosigenes). When a word occurs no more than ten times, its occurrences are listed; in most other cases only the number of occurrences is given. Occurrences in lemmas are ignored. Sometimes comparatives, superlatives, and adverbs are included under the positive form of an adjective, sometimes they are treated separately. There is a separate index of names.

Index of Passages

(a) Testimonia and fragments

I list here passages from Simplicius which have been collected as testimonia about or fragments of various ancient authors.

(b) Texts quoted, paraphrased, or clearly referred to by Simplicius or Alexander

(c) Index of early texts cited in the notes

References are to the line in the Greek text on which a footnote number occurs.

Index of Names

In many cases information or references to where information can be found are provided in the notes on a given passage. For sun and moon see the English-Greek Glossary and the Greek-English index; see also the Introduction. For Alexander, where possible I indicate Simplicius' assessment of what Alexander says. For Aristotle I list passages in which something general is said about Aristotle or there is a reference to material outside *On the Heavens*.

(a) Astronomical names

(b) Individuals and groups

(c) Modern scholars

Subject Index

This index lists places where Simplicius' discussion goes beyond straightforward exposition of Aristotle's text. See also the other indices and the English-Greek Glossary.